The History of Western Palearctic Falcons

Giovanni Leonardi

The History of Western Palearctic Falcons

Springer

Giovanni Leonardi
Hierofalcon Research Group
Catania, Italy

ISBN 978-3-031-86527-5 ISBN 978-3-031-86528-2 (eBook)
https://doi.org/10.1007/978-3-031-86528-2

This Springer imprint is published by the registered company Springer Nature Switzerland AG
The registered company address is: Gewerbestrasse 11, 6330 Cham, Switzerland

To my mother

–Giovanni Leonardi

Preface

This seems to be making the evidence fit the verdict and not the verdict the evidence, which may do for art-history but hardly for the history of natural history. (Yapp 1982)

The excerpt above, written by the zoologist William Brunsdon Yapp (1909–1990), encapsulates the methodological approach employed in this book. Specifically, Yapp disagreed with the iconographic interpretation of a 700 CE carving from Cumberland that portrays a falconer and his bird (Yapp 1982). From a zoological perspective, there was no element that could identify the falconer as Saint John because the bird was not remotely an eagle, and the falconer's stick was not at all a pen (Yapp 1982). A significant part of this book involves interpreting historical sources such as the carving from Cumberland. Unfortunately, the way historians and ornithologists interpret these sources can vary greatly. While I'm not a historian myself, my approach aligns with Yapp's. Nonetheless, another essential aspect of this work is the necessity of a multidisciplinary approach. This is crucial for reliably understanding the past distribution and population sizes of falcons in the Western Palearctic ecozone. For instance, the name given to a falcon in a medieval falconry manual may not correspond to the species we recognize today (see Chap. 4). This field of study by historians and linguists offers insight through etymology and the analysis of past prose and poetry. It is an extremely complex approach, with Dafydd Evans' text on the name "*lanier*" (= lanner falcon) being an outstanding example (Evans 1967). However, it is important to be cautious when assuming species distribution solely based on their names (see Chap. 5). Otherwise, we might mistakenly believe that sub-tropical species were common in colder areas, suggesting a warmer past climate. This issue is similar to the identification of Saint John, as previously described by Yapp. The most effective approach seems to vary depending on the species. It is relatively reliable with widespread species such as the common kestrel or the Eurasian hobby, but considerably more challenging with rarer ones like the lanner or saker falcons. Although widely regarded as a comprehensive source on falcons in the Middle Ages, Frederick II's "De arte venandi cum avibus" includes controversial illustrations and text on rare species (Fig. 1; see Sect. 4.3.1; Yapp 1983). Nevertheless, throughout written history, there are many accurate examples

Fig. 1 Falcons and deer depicted in the *De arte venandi cum avibus* (1240s) MS Pal. lat. 1071 folio 49 recto (Credit: Wikipedia) Most of the falcons are likely intended to represent peregrine falcons. Any differences in plumage color, markings, or eye color appear random and unintentional. (Yapp 1983)

that remain relevant today. For instance, in the natural encyclopedia "Liber de natura rerum" by Thomas of Cantimpré around 1240, cooperative hunting by falcons is described with remarkable accuracy: *In captura vero ardee cum duo falcones pariter relaxantur, socialiter volant. Unus summa petit, alius iuxta terram in imo volat, videlicet ut ille qui in summa contendit ardeam volantem impulsu precipitet, et ille qui in imo volat precipitatam accipiat* (When hunting herons, two falcons are released together and fly together. One flies high to scare the heron into flight, while the other flies low to catch it after it descends; Böse 1973). Ultimately, a careful multidisciplinary approach is strongly advisable.

Fossils and sub-fossils provide the most reliable insight into the historical distribution and population sizes of falcons (Fig. 2; Yalden and Albarella 2009). In general, their occurrence in fossil assemblages across archaeological sites in the Western Palearctic broadly reflects their past distribution and abundance in the field (Fig. 2; Tyrberg 2008; Kessler 2014). Additionally, these remains help identify extinction events that occurred during major shifts between geological periods (Tomek et al. 2012; Kessler 2018). Even in their sub-fossil form, they provide substantial support for other historical sources such as written records and artifacts (Ericson and Tyrberg 2004).

Considering the total time span of roughly 1,800,000 years from the Early Pleistocene to the Neolithic period, the occurrence coefficients of different falcon species in fossil assemblages clearly reveal their intrinsic differences (Fig. 2). Despite spanning various time periods, remains of common species like the common kestrel and the peregrine falcon are widely found in archaeological sites (Fig. 2). Similarly, species that are currently uncommon or localized are represented by low percentages in past assemblages (Fig. 2). This observation suggests that there have been minimal changes in the distribution of Western Palearctic falcons, except on a local scale (see Chaps. 1 and 2). Furthermore, it is notable that no species of the genus *Falco* has become extinct over this extended period, except for

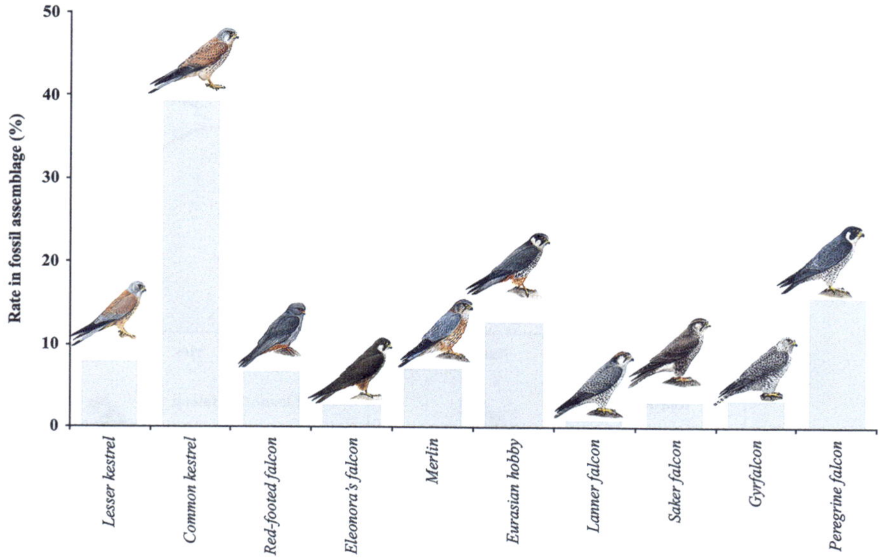

Fig. 2 Pooled rates in fossil assemblages of falcon species from the Early Pleistocene (1,800,000 years BCE) and the Neolithic period (5500 years BCE; drawings courtesy of Marco Preziosi; the data are included in the appendices of Chaps. 1 and 2)

presumed subspecies (see Sect. 1.4 and Table 2.3). Accordingly, modern Palearctic species mostly appeared during the Pliocene and Early Pleistocene, with very few emerging thereafter (Finlayson 2011).

Climate plays a crucial role in determining the distribution of bird species on a broad scale, including falcons (Wiens and Donoghue 2004; Popescu et al. 2024). Data shown in Fig. 2 suggest that falcon species exhibit niche conservatism, meaning they retain ancestral ecological characteristics and have difficulty adapting to new conditions (Wiens and Graham 2005). This conservatism primarily restricts their geographical range (Wiens and Donoghue 2004; Wiens and Graham 2005). A comprehensive study of the climatic niche of the genus *Falco* in Europe aligns with these findings (Fig. 2; Popescu et al. 2024). Among European falcons, the common kestrel, peregrine falcon, and Eurasian hobby exhibit the largest climatic niches (Fig. 2; Popescu et al. 2024). In contrast, the red-footed falcon, merlin, and gyrfalcon are situated at intermediate positions in terms of climatic specialization (Fig. 2; Popescu et al. 2024). The lesser kestrel, lanner falcon, and Eleonora's falcon are the most specialized species, with their habitats primarily located in southern latitudes (Fig. 2; Popescu et al. 2024). The saker falcon also has a restricted climatic niche, but unlike the previously mentioned subtropical species, it is found in mid-latitude areas (Fig. 2; Popescu et al. 2024).

Ultimately, observations from additional historical sources should be interpreted within the ecological framework inherent to each particular species (Figs. 3 and 4). Species that can adapt to mutated environmental conditions are likely to see

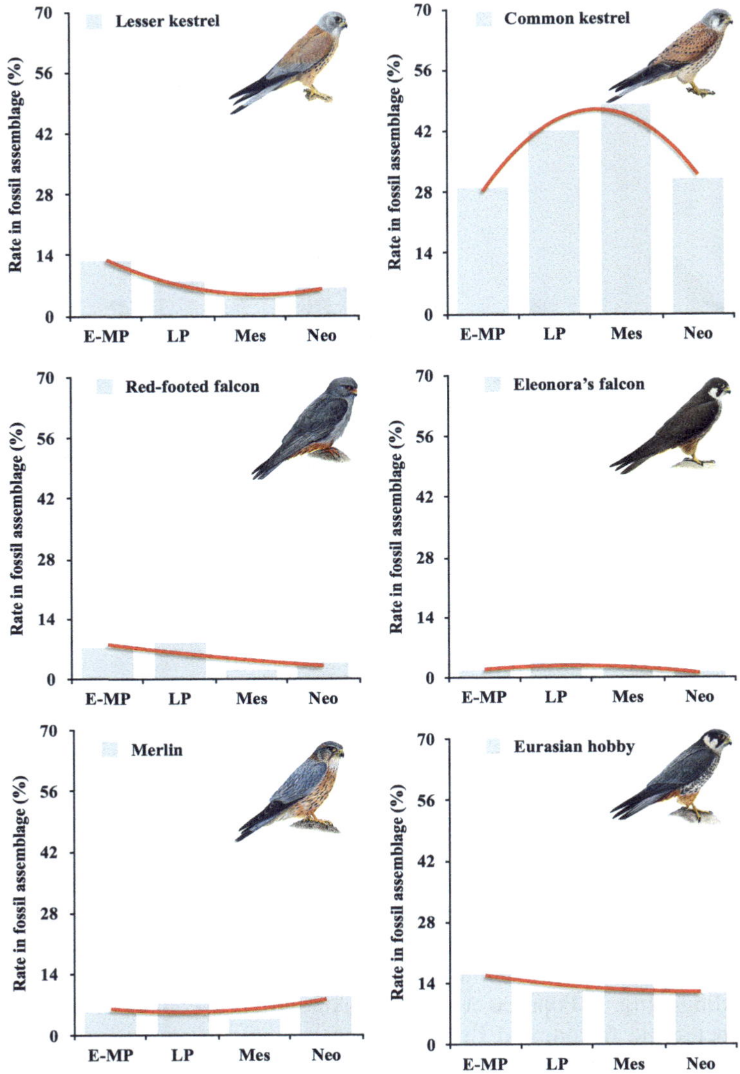

Fig. 3 Percentage rates in fossil assemblages of small falcon species from the Early-Middle Pleistocene to the Neolithic period (*E-MP* Early and Middle Pleistocene; *LP* Late Pleistocene; *Mes* Mesolithic; *Neo* Neolithic; drawings courtesy of Marco Preziosi; the data are included in the appendices of Chaps. 1 and 2)

long-term increases (Figs. 3 and 4). Conversely, highly specialized species may retain some resilience, but are at risk declines over time (Figs. 3 and 4; Huntley et al. 2007; Lavergne et al. 2013). It is also essential to evaluate the increasing impact of human activities on falcon populations. This includes exploitation by Neanderthals

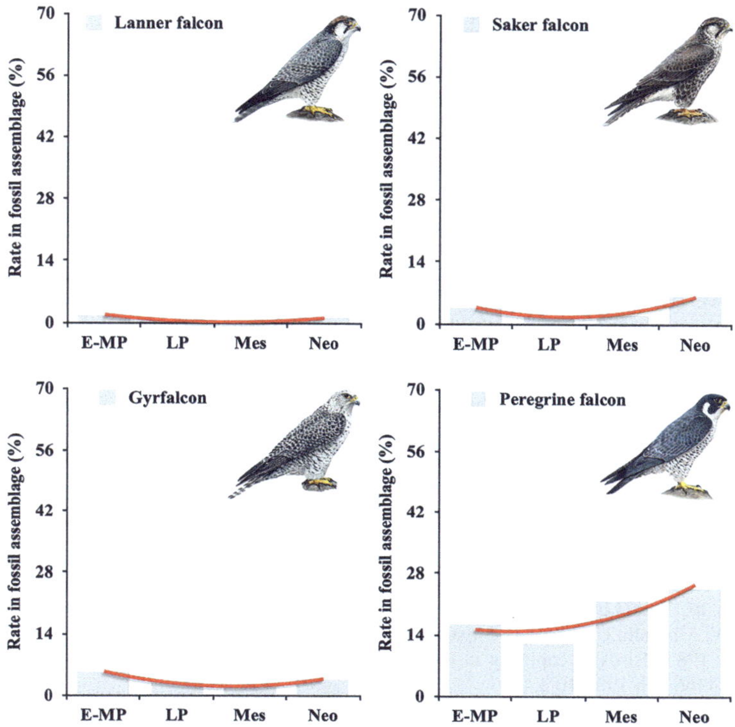

Fig. 4 Percentage rates in fossil assemblages of large falcon species from the Early-Middle Pleistocene to the Neolithic period (*E-MP* Early and Middle Pleistocene, *LP* Late Pleistocene; *Mes* Mesolithic; *Neo* Neolithic; drawings courtesy of Marco Preziosi; the data are included in the appendices of Chaps. 1 and 2)

(Chap. 1), captures for rituals or falconry (Chaps. 3, 4, and 5), direct destruction as pests (Chaps. 5 and 6), and indirect effects from chemicals used in modern agriculture (Chap. 7).

The growing human pressures, including global warming, poses significant challenges for falcons in the near future. An integrated, multidisciplinary approach to understanding past population changes can provide valuable insights with implications beyond historical contexts.

Catania, Italy Giovanni Leonardi
6 May 2024

References

Böse H (1973) Thomas of Cantimpré, Liber de natura rerum. De Gruyter, Berlin

Ericson PGE, Tyrberg T (2004) The early history of the Swedish avifauna: a review of the subfossil record and early written sources. Kungliga Vitterhets Historie och Antikvitets Akademiens Handlingar, Antikvariska Serien 45:1–349

Evans D (1967) Lanier, l'histoire d'un mot. Librairie Droz, Genève

Finlayson C (2011) Avian Survivors. T & AD Poyser, London

Huntley B, Gree RE, Collingham YC, Willis SGA (2007) Climatic atlas of European breeding birds. Lynx Edicions, Barcelona, Spain

Kessler J (2014) Fossil and subfossil bird remains and faunas from the Carpathian Basin. Ornis Hung 22:65–125. https://doi.org/10.2478/orhu-2014-0019

Kessler J (2018) Evolution and presence of diurnal predatory birds in the Carpathian Basin. Ornis Hung 26:102–123. https://doi.org/10.1515/orhu-2018-0008

Lavergne S, Evans MEK, Burfield IJ, Jiguet F, Thuiller W (2013) Are species' responses to global change predicted by past niche evolution? Philos Trans R Soc B 368:20120091. https://doi.org/10.1098/rstb.2012.0091

Popescu SM, Tigae C, Dobrițescu A, Ştefănescu DM (2024) Exploring the climatic niche evolution of the genus *Falco* (*Aves*: *Falconidae*) in Europe. Biology 13:113. https://doi.org/10.3390/biology13020113

Tomek T, Bocheński ZM, Socha P, Stefaniak K (2012) Continuous 300,000-year fossil record: changes in the ornithofauna of Biśnik Cave, Poland. Palaeontol Electron 15:2A–20p. https://doi.org/10.26879/277

Tyrberg T (2008) Pleistocene birds of the Palearctic. http://web.telia.com/~u11502098/pleistocene.pdf

Wiens JJ, Donoghue MJ (2004) Historical biogeography, ecology and species richness. Trend Ecol Evol 19:639–644. https://doi.org/10.1016/j.tree.2004.09.011

Wiens JJ, Graham CH (2005) Niche conservatism: integrating evolution, ecology, and conservation biology. Ann Rev Ecol Evol Syst 36:519–539. https://doi.org/10.1146/annurev.ecolsys.36.102803.095431

Yalden DW, Albarella U (2009) The history of British birds. Oxford University Press, Oxford

Yapp WB (1982) Birds in captivity in the Middle Ages. Arch Nat Hist 10:479–500. https://doi.org/10.3366/anh.1982.10.3.479

Yapp WB (1983) The illustrations of birds in the Vatican manuscript of *De arte venandi cum avibus* of Frederick II. Ann Sci 40:597–634. https://doi.org/10.1080/00033798300200401

Acknowledgments

This book could not have been prepared without the kind assistance of many people and institutions. This project was initiated around ten years ago, and numerous individuals have generously contributed their personal observations, publications, suggestions, and comments.

I am especially thankful to Marco Preziosi for granting permission to use their illustrations to complement the maps and diagrams.

Contents

Chapter 1
Pleistocene

1.1 Prior to the Pleistocene

In the early Miocene (ca. 23–16 Mya), the *Falconidae* family rapidly spread from South America to Eurasia and Africa via North America (Griffiths 1999; Finlayson 2011; Fuchs et al. 2015; Cenizo et al. 2016). During this epoch, temperate forests, grasslands, and steppes greatly benefitted from the global shift from subtropical to cooler climates (Blondel and Mourer-Chauviré 1998). While an African origin has often been suggested, an early fossil from the Lower Eocene (56–33.9 Mya) found in Antarctica supports the hypothesis of a Neotropical or Austral origin for this group (Olson 1985; Cenizo et al. 2016).

The *Falconinae* subfamily began diversifying between 12.6 and 19.3 Mya, and the divergence of the genus *Falco* was favored by the spread of savannahs during the late Miocene (Fuchs et al. 2015). True falcons separated from *Polihierax* and *Microhierax* genera around 10.1 Mya and 15.3 Mya by increasing aridity after the mid-Miocene Climatic Optimum. This differentiation coincided with a series of strong, short-term, wet–dry and cool–warm climatic fluctuations (Blondel and Mourer-Chauviré 1998). However, after the initial burst of diversification, the process slowed. Consequently, members of this genus exhibit very low genetic divergence despite geographical segregation and current systematic classifications (Nittinger et al. 2005, 2007; Fuchs et al. 2015; Yang et al. 2018). During the Plio-Pleistocene glaciations, when climate becomes much colder, speciation could occur due to the geographical separation of populations, which occasionally overlapped and allowed hybridization (Finlayson 2011). Presumably, varying degrees of isolation, due to an increase in open habitats within tertiary forests, could have led to a shift from well-defined species to analogous species and differentiation at the subspecies level (Finlayson 2011).

G. Leonardi, *The History of Western Palearctic Falcons*,
https://doi.org/10.1007/978-3-031-86528-2_1

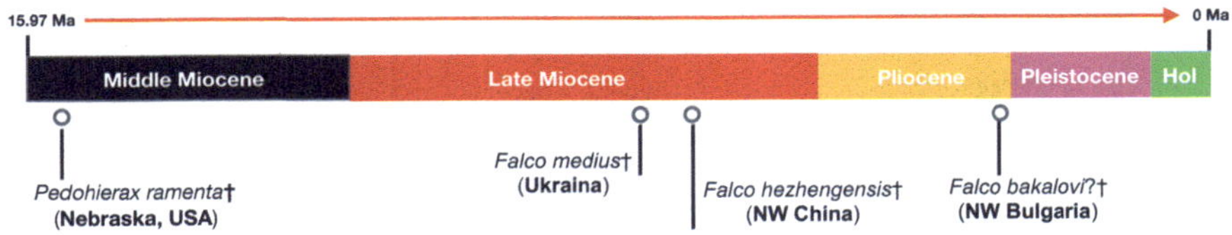

Fig. 1.1 Fossils of the genus *Falco* from before the Pleistocene epoch, as reported by Umanskaya (1981), Becker (1987), Boev (1999), Mlíkovský (2002), and Li et al. (2014)

Unfortunately, there is a scarcity of well-preserved pre-Pleistocene fossils in Eurasia and none in Africa (Fig. 1.1; Groombridge et al. 2002; Li et al. 2014). It is likely that the Late Miocene avifaunas of Eurasia have not been consistently studied, and many forms assigned to species can only be reasonably assigned to families (Zelenkov 2017). Large falcons (i.e., *Hierofalco* subgenus) probably originated in Africa around the Late Pliocene (2.5–2 Mya; Nittinger et al. 2005). This group is relatively young compared to kestrels, which include an ancestor such as *F. medius* from the late Miocene (Fig. 1.1; Mlíkovský 2002). A new kestrel-like fossil from late Miocene of north-eastern China (*F. hezhengensis*) supports the Eurasian origin of kestrels (Fig. 1.1; Li et al. 2014).

From the late Miocene to the early and middle Pliocene, kestrels, hobbies, and other larger true falcons radiated in the Western Palearctic, as confirmed by fossils. Globally climatically tolerant falcons, such as the common kestrel, the Eurasian hobby, and the peregrine falcon, became widespread. Another group including lanner, sooty, barbary, and saker falcons occupied driest areas, while a few others, such as gyrfalcon and merlins, adapted to cold climates (Finlayson 2011).

Recent analyses using the complete mitochondrial genome of several falcon species have confirmed that common and lesser kestrels were the first to emerge, followed by merlins and a clade that includes peregrine falcons and hierofalcons (Zhan et al. 2013; Sveinsdóttir et al. 2017; Doyle et al. 2018). The split between the saker and peregrine falcons is estimated to have occurred about 2.1 Mya (Zhan et al. 2013). However, genomic maps reveal substantial similarities between peregrine and saker falcons, with few inter- and intrachromosomal rearrangements and complete synteny between the latter and the gyrfalcon (Joseph et al. 2018).

1.2 Pleistocene Palaeoclimatic Conditions in the Western Palearctic

The Pleistocene is the first of two epochs that comprise the Quaternary period (Table 1.1). This epoch is characterized by climatic oscillations and a succession of glacial and interglacial periods during which vast continental ice sheets repeatedly advanced and receded (Holman 1998). In the Western Palearctic ecozone, these ice sheet movements significantly impacted the distribution of plant and animal life, as extensive habitats were cyclically covered and uncovered (see also Sect. 1.3;

Table 1.1 Timetable of Pleistocene stages (Ma = mega annum, ka = kilo annum)

Time period	Age
Gelasian (age of the formerly Early Pleistocene)	2,58 Ma–1,80 Ma
Calabrian (age of the formerly Early Pleistocene)	1,80 Ma–774 ka
Chibanian (formerly Middle Pleistocene)	774 ka–129 ka
Late Pleistocene (Tarantian)	129 ka–11,700

Table 1.2 Timetable of the last interglacial and last glacial cycle (ka = kilo annum)

Time period	Age	MIS stage
Eemian Interglacial	123 ka	MIS 5e
Weichsel Early Glacial	115 ka–60 ka	
Weichselian High Glacial	60 ka–15 ka	
Last Glacial Maximum (LGM)	22 ka	MIS 2
Weichselian Late Glacial	15 ka–9,6 ka	

Holman 1998; Holm and Svenning 2014). The Late Pleistocene in the Western Palearctic has received a great deal of attention for its climatic changes and their effects on biotic dynamics (Holm and Svenning 2014). This period comprises the Last Eemian (last interglacial) and the Weichselian (last glacial), ending with the onset of the current interglacial, the Holocene (Table 1.2).

1.2.1 The Last Glaciation Cycle

The Weichselian glaciation of northern and central Europe, equivalent to the Würm glaciation of the Alpine zone, refers to the last glacial cycle. This cold stage is characterized by peaks in global sea level and marine oxygen isotope records (MIS), encompassing nearly the entire Late Pleistocene (Table 1.2).

The Eemian, the last interglacial period before the Holocene (MIS 5e), experienced a warmer climate (more than 2° above present) and higher sea level (>4–6 m above current levels; McCarty 2001). Forests predominantly covered regions reaching far northern localities, and the Greenland ice sheet was greatly reduced (McCarty 2001; Holm and Svenning 2014). Starting from the MIS 5d, temperatures declined significantly, though interglacial stages 5c and 5a were warmer. The global ice cover reached its maximum during the Last Glacial Maximum (LGM; ca. 22,000 years ago), and this glacial cycle concluded with the Younger Dryas (Table 1.2; Fig. 1.2; Hughes et al. 2013).

During the Weichselian Stage, the Fennoscandinavian ice sheet was one of the major global ice masses, merging with the British–Irish ice sheet and reaching its southernmost limit in central Poland (Graham et al. 2011; Clark et al. 2012; Hughes et al. 2013). The southwestern sector of the Fennoscandinavian ice sheet reached its maximum extent during MIS 4 and 3, whereas the south-eastern sector's extent was

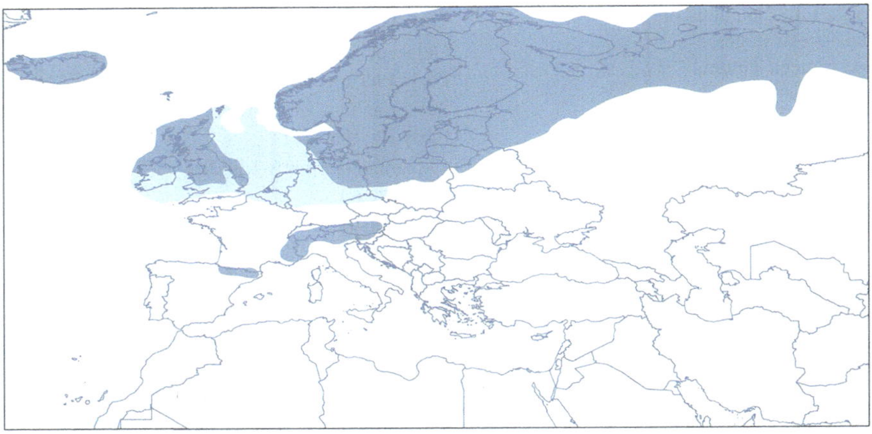

Fig. 1.2 Distribution of the ice sheet (grayish azure) and Polar and Alpine deserts (cyan-blue) during the Last Glacial Maximum (25–15 ka BP; redrawn from Ray and Adams 2001)

synchronous with the global record of ice volume changes (Houmark-Nielsen 2011; Hughes et al. 2013). Similarly, the maximum extent of glaciers in the Pyrenees precedes the global LGM, while in the Alps, it was almost synchronous with the global LGM (Hughes et al. 2013).

1.2.2 Vegetation and Habitats

Reconstructing the exact range of plants during the Pleistocene is challenging. However, a basic understanding of their historical distribution can be achieved by examining their tolerance to minimal mean annual temperatures (Hardy 2010). Generally, vegetation composition varied along north–south and east–west gradients (Finlayson 2011). Warm, marine temperatures supported coniferous and deciduous forests, while cold, continental conditions favored open sub-Arctic steppe-tundra.

The Early Pleistocene landscape was characterized by a series of steppe, evergreen-bushland, and mixed-forest biomes (Altolaguirre et al. 2020). Increased precipitation during both glacial and interglacial periods likely facilitated the formation of fully wooded habitats (Altolaguirre et al. 2020). The late Middle Pleistocene experienced significant climatic fluctuations, with numerous glacial–interglacial cycles (Urban et al. 2011). During glaciations, open and dry vegetation communities emerged in southern latitudes of both the Western and Eastern Mediterranean regions (Fig. 1.3). These landscapes were predominantly steppes, or, if the climate was sufficiently humid, wooded steppes (i.e., southern Italy; Yll et al. 2006; Altolaguirre et al. 2020). Conversely, climate change during interglacial periods might have allowed forested ecosystems to exist while maintaining some openness (Altolaguirre et al. 2020).

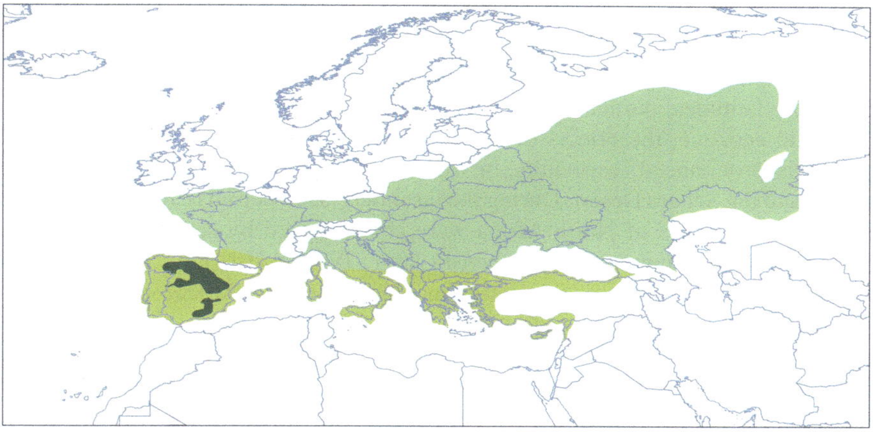

Fig. 1.3 Palearctic and Asia LGM (Last Glacial Maximum) vegetation map. Steppe–tundra (light green) was a widespread vegetation type, combining plants from both tundra and steppe ecotypes (ca. 50% ground cover by plants and the rest patchy bare). Forest steppe (light green) features 5–20% tree cover, typically in patches within a predominantly grassy landscape. Tundra ecotype is indicated by dark green. (Redrawn from Ray and Adams 2001)

Avian fossil assemblages indicate the presence of trees north of the Alps even during full-glacial stages (Fig. 1.3). Evidence suggests that Central Europe was either forested or had significant forest patches at the end of the Pleistocene (Núñez-Lahuerta et al. 2018).

1.2.3 Glacial and Interglacial Refugia

During glacial periods, refugia for temperate species were located in Spain, southern Italy, Sicily, the southern Balkans, the southern Caucasus, and along the northern border of the Caspian Sea (Taberlet et al. 1998; Sánchez-Marco 2004; Murray et al. 2010). Bird behavior in the past was presumably different from today, with nesting areas likely situated further south during colder climates (Jenkinson et al. 1984). Nevertheless, species have responded individually to increases in habitat availability caused by climatic changes at the end of the last glaciation. In these regions, most bird populations, including warm-climate species such as Eleonora's falcon, the lesser kestrel, and the lanner falcon, have survived (Blondel 1985; Tyrberg 2008a).

During the Late Quaternary, cold-adapted species experienced recurring patterns of glacial expansions and interglacial contractions (Stewart et al. 2010). In this context, predator dispersal was driven by prey availability and movement (Dalén et al. 2005; Nyström et al. 2006). For example, the gyrfalcon maintained close and long-standing relationships with its principal food sources, such as Rock Ptarmigan (*Lagopus mutus*) and Arctic hares (*Lepus* ssp.; Holder et al. 1999; Waltari and Cook 2005). This large falcon likely followed these prey populations into new open steppe–tundra environments created by glacial retreat, where fewer potential prey survived (Núñez-Lahuerta et al. 2018). However, like other Arctic or steppe species, no relict populations persisted from the glacial range expansion (Tyrberg 1991).

At low latitudes, particularly in the Afrotropics and, to a lesser extent, in the Palearctic ecozone, the effects of Pleistocene glaciations on species distribution were less pronounced, especially in regions like Mediterranean Europe, North Africa, and the Sahara Desert. During the last interglacial period (LIG), for instance, the hyper-arid zones of North Africa were characterized by savannah grasslands with shallow, permanent lakes (Bocheński 1991; Tyrberg 2010). As a result, fossilized avian assemblages from this period include many Afrotropical species that are still found in the Sahel savannah today, as well as some species that inhabit the semi-desert areas at the northern edge of the Sahara Desert (Bocheński 1991; Tyrberg 2010). Among falcons, this climatic shift likely led to the concentration of lanner falcon populations in restricted breeding areas across the Maghreb, Mediterranean Africa, and the Sahel (Leonardi 2015).

1.3 Fossil Assemblages

The fossil record primarily consists of bone fragments that reveal specific traits of past environments (Winn Jones 2006). Extant species serve as valuable indicators for reconstructing palaeoenvironments because their ecological requirements are well documented in ornithological literature (Sánchez-Marco 2004; Finlayson et al. 2011). Consequently, the composition of fossilized avian assemblages is believed to be affected by palaeoclimatic conditions (Sánchez-Marco 2004; Stewart et al. 2010). However, in Pleistocene assemblages, bird species that currently inhabit different environments often appear together (see Sect. 1.2.3; Tyrberg 1991). Additionally, a comprehensive approach on past avian communities is usually hampered by the incomplete representativeness of the accessible fossil assemblages and the presence of extinct species (Guerra et al. 2013).

Although a direct link to climate should be plausible, avian assemblages primarily reflect changes in vegetation. As a result, cold- and dry-adapted species dominated Late Pleistocene fossil assemblages (Holm and Svenning 2014). During glaciations, northern species shifted southward, while open-vegetation species moved northward and westward from the south and east, respectively (Holm and Svenning 2014). For example, several species currently found in Arctic or sub-Arctic regions, such as the gyrfalcon, were typical of past cold environments (Fig. 1.4; Stewart et al. 2010). Similarly, the red-footed falcon exhibited longitudinal range shifts, with a notable preference for westward expansion compared to its present distribution (Fig. 1.4; Holm and Svenning 2014).

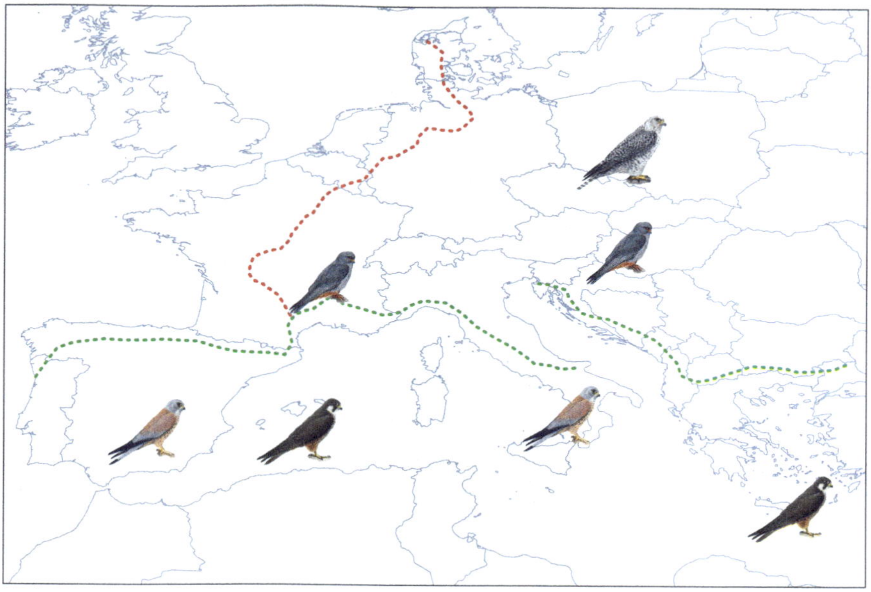

Fig. 1.4 Distribution of some falcon species based on fossil remains during the LGM. The red line indicates the present boundary between the Atlantic and Central European territories, while the green dotted line marks the current Eurosiberian–Mediterranean boundary. Warm-climate species (e.g., lesser kestrels and Eleonora's falcons) occupied the Mediterranean basin as they do today. In contrast, the gyrfalcon exhibited a significant latitudinal range shift, and the red-footed falcon showed a westward expansion. (Redrawn from Ray and Adams 2001; Hardy 2010; Holm and Svenning 2014; Núñez-Lahuerta et al. 2018; Badino et al. 2020; Di Pasquale et al. 2020; data from Appendix; drawings courtesy of Marco Preziosi)

1.4 Extinct Species and Subspecies

In the West Palearctic subregion, moderate extinctions occurred during the Late Pleistocene, primarily affecting scavengers (Tyrberg 2008b). Fossil evidence from early and middle Pleistocene deposits includes presumed extinct falcon species and subspecies (Fig. 1.9; Jánossy 1972; Mourer-Chauviré 1975; Louchart 2002). These ancient falcons have raised numerous questions about their apparent differences from extant species and their ecological roles (Kessler 2018). Zelenkov (2013) contends that classifying a fossil as a subspecies of an extant species is conceptually problematic.

Jánossy (1972) described *F. tinnunculus atavus*, a subspecies of the common kestrel that lived during the early and middle Pleistocene and possibly before that epoch (Fig. 1.5; Mlíkovský 2002; Kessler 2009). However, recent revisions have excluded this subspecies from Pliocene (Polgárdi, Hungary) and Middle Pleistocene (Hundsheim, Austria) assemblages (Mlíkovský 2009; Zelenkov 2017). Despite this, the fossilized remains of *F. tinnunculus atavus* are slightly larger than those of the

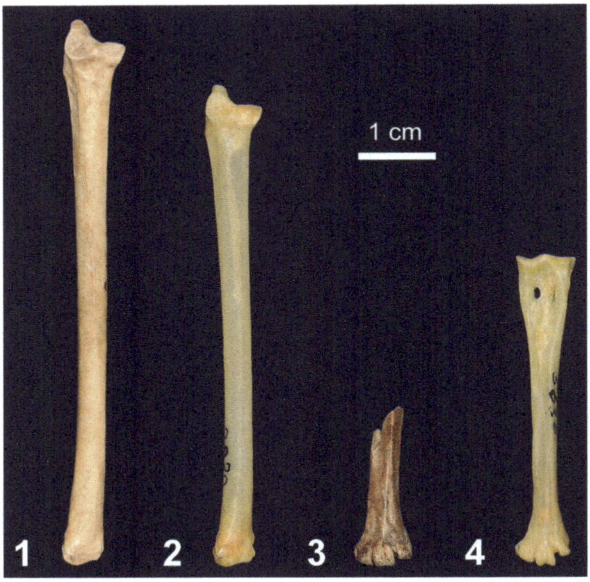

Fig. 1.5 Left ulna in cranial view and left tarsometatarsus in dorsal view (1 and 3: *F. tinnunculus atavus*; 2 and 4: *F t. tinnunculus*). (Credit: Zbigniew M. Bocheński)

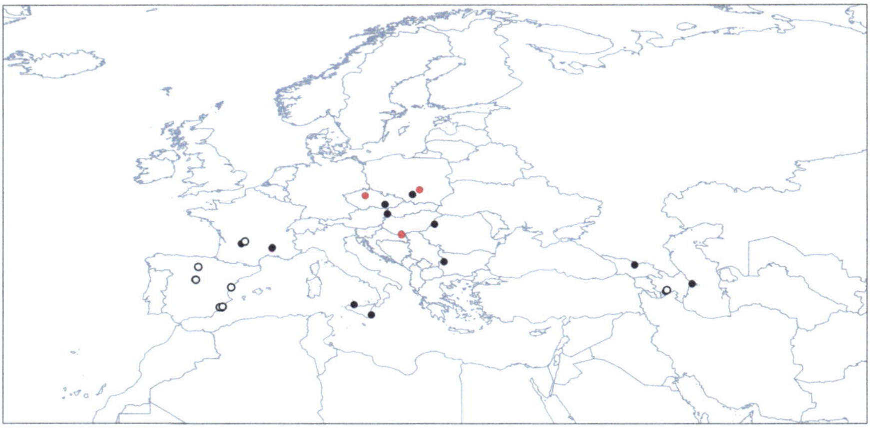

Fig. 1.6 Archaeological sites with fossilized remains of *F. t. atavus* (red dots), common kestrel (black dots), and lesser kestrels (white dots) from the Early and Middle Pleistocene (see Appendix for data). The coexistence of this extinct subspecies with the nominate species in Central Eastern Europe appears plausible

modern common kestrel (Fig. 1.5; Jánossy 1972; Mourer-Chauviré 1975, 1993; Tomek et al. 2012). Additionally, Tomek et al. (2012) proposed that this extinct subspecies may have coexisted with the nominate species (Fig. 1.6). The kestrel-like "*tinnunculus*" group is generally considered the sister group to the larger greater kestrel (*F. rupicoloides*) of southern Africa, both descending from a common

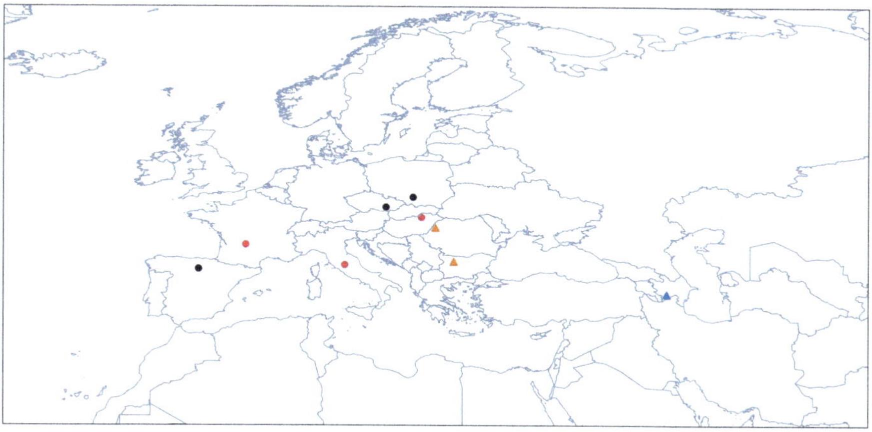

Fig. 1.7 Archaeological sites with fossilized remains of *F. antiquus* (red dots), gyrfalcons (black dots), sakers (yellow triangles), and lanner falcons (blue triangle) from the Early and Middle Pleistocene. (see Appendix for data)

ancestor (Wink and Sauer-Gürth 2004). The larger size of *F. t. atavus* is thought to represent a primitive trait that persisted until the end of the Middle Pleistocene. Conversely, unique morphological features may reflect adaptations of certain populations to specific ecological conditions that were geographically and temporally restricted (Mourer-Chauviré 1993).

Similarly, *F. antiquus* has been proposed as an intermediate form between the saker falcon and the gyrfalcon, potential serving as their common ancestor (Mourer-Chauviré 1975; Potapov and Sale 2005; Johnson et al. 2007). Pfeffer (2012) suggests it may represent an ancient, lanner-like ancestor. However, Mlikovskỳ (2002) argued that there are no significant morphological differences between *F. antiquus* and the modern saker falcon. Unfortunately, this species has been identified in only few archaeological sites in Southern Europe (France, Hungary, and Italy) and is thought to have expanded eastward through Asia into North America (Fig. 1.7; Janossy 1977; Piperno et al. 1984; Mourer-Chauviré et al. 2003; Potapov and Sale 2005; Johnson et al. 2007). Changes in paleoclimate after the last ice age led to the formation of a large forest belt that fragmented and isolated populations of the common ancestor, thereby promoting the speciation (Karyakin 2011; Pfeffer 2012). As a result, saker falcon populations are defined by two main genetic clusters (eastern and western), with an intermediate third cluster in Central Eurasia situated between the two main groups (Zhan et al. 2015). Additionally, as an adaptation to cold open habitats, the saker falcon displays genes associated with hypoxia and immune response (Zhan et al. 2013; Pan et al. 2017).

1.5 Human Exploitation of Falcons

Falcons most likely roosted or nested in caves and died naturally there, though some were prey of human or animal predators. Overall, prehistoric hunter-gatherers chose large mammals as their main food source (Laroulandie 2009; Richards and Trinkaus 2009). Unlike birds, large mammals required less hunting effort and provided the highest rate of energy intake (Speth and Spielmann 1983; Laroulandie 2009). While birds could be exploited as alternate prey, their feathers and bones were also used as raw materials for crafts or symbolic purposes (Laroulandie 2009; Morin and Laroulandie 2016). The presence or absence of falcon species in an assemblage is presumably determined primarily by their geographical proximity to human populations. However, their preferences for open habitats such as steppes and grasslands effectively separated them from human settlements (Finlayson 2019).

Unfortunately, considerable difficulties arise in distinguishing within the species assemblage between bones accumulated through the activity of predators, scavengers, or humans (Laroulandie 2002; Bochenski et al. 2009). Nevertheless, peculiar marks on bones made by teeth, beaks, claws, and gastric juices reveal predation or scavenging by nonhuman agents (Laroulandie 2005). Thus, fossilized falcon bones found in human settlement may not be exclusively related to their exploitation or use (Rufà and Laroulandie 2021). During the Last Glacial Period (LGP), but also at a Neanderthal site, de-fleshing activities on falcon bones are evident through the striae observed on their surface (see Sect. 1.5.1; Martínez Valle et al. 2016; Rufà and Laroulandie 2021). On the other hand, falcons *per se* are agents of accumulation, influencing the assemblage composition as well as the occurrence of bone remains (Huguet et al. 2017).

1.5.1 Neanderthals

Different falcon species were found in Neanderthal (*Homo neanderthalensis*) settlements across Europe (Peresani et al. 2011; Lloveras et al. 2018; Blanco et al. 2021). Several of these sites were primarily carnivore dens (e.g., hyenas) that were periodically used by small groups of Neanderthals during their travels, most likely during warmer periods of the Last Glacial. Neanderthals opportunistically captured small falcons (common kestrels, lesser kestrels, and red-footed falcons) while exploiting other species (e.g., coughs *Pyrrhocorax* spp.; Peresani et al. 2011; Blanco et al. 2021). All these species are associated with rock cliffs, caves, and open ecotones near forested zones (Peresani et al. 2011). A recent review confirmed the prevalence of common kestrels in the majority of Neanderthal settlements, with lesser kestrels and red-footed falcons being much less frequent, and gyrfalcons, Eleonora's falcon, peregrines, and merlins rarely found (Finlayson 2019).

Given their lack of food utility, falcon skeletons, especially wings, were presumably used for symbolic purposes, such as extracting feathers or raptor claws

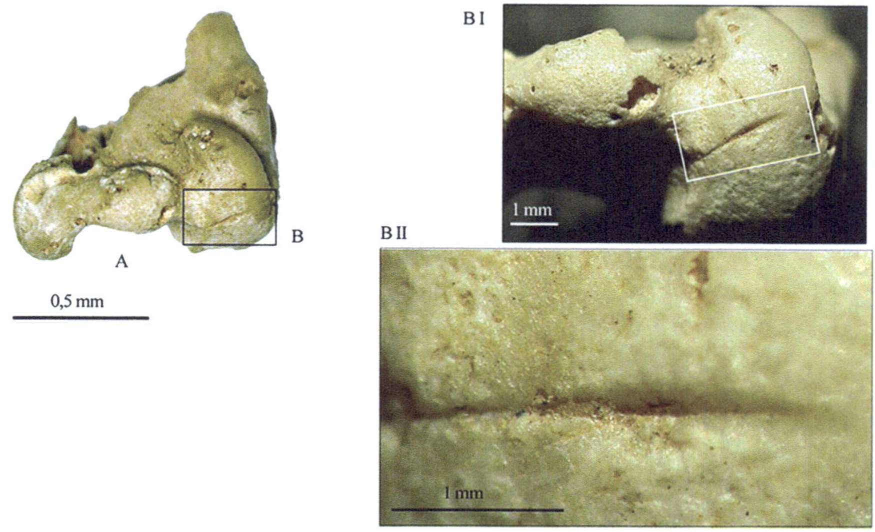

Fig. 1.8 Cut marks on the left distal humerus of a red-footed falcon (*Falco vespertinus*). Localization of the cut marks: BI and B II. (Credit: Marco Peresani)

(Peresani et al. 2011; Finlayson et al. 2012a, b; Morin and Laroulandie 2016). Overall, striae, crushing marks, and peeling produced by lithic artifacts are primarily related to the disarticulation of humerus from the ulna (Fig. 1.8; Peresani et al. 2011; Oros Sršen et al. 2014; Lloveras et al. 2018). Indeed, these marks are frequently linked to meat consumption (Lloveras et al. 2018).

1.6 Western Palearctic Falcons

The coexistence of ecologically distinct and similar falcon species in the same territory depends on their habitat, breeding, and feeding requirements (Leonardi 2020). Niche partitioning and the reduction of competition among similar species can enable their presence within the same guild (Pande et al. 2018; Berlusconi et al. 2022). Specialist species, which are closely tied to specific eco-climatic conditions, often exhibit high spatial correlation and restricted ranges (Berlusconi et al. 2022). In contrast, generalist species, such as the common kestrel, are widespread due to their ability to adapt to a broad range of climates (see also the "Preface").

Recent findings show that the Pleistocene climate cooling had a profound impact on the composition and structure of geographical species pools, often more so than long-time evolutionary processes (Calatayud et al. 2019 and references therein). The past distributions of falcons reveal habitat requirements similar to those of today, although some species seem to have been associated with different regions during the Pleistocene (see Fig. 1.4; Sánchez-Marco 1999; Holm and Svenning,

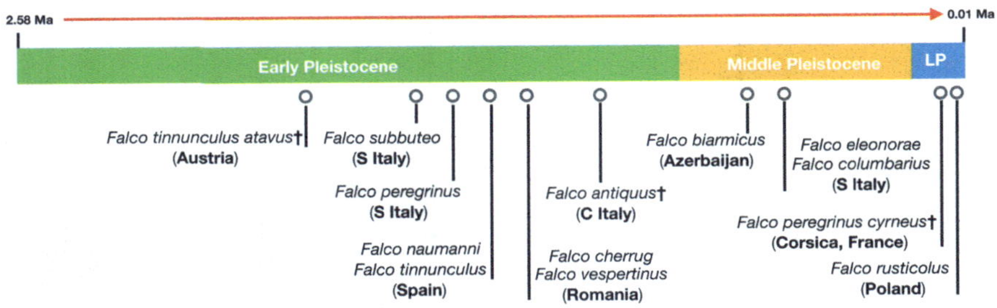

Fig. 1.9 Fossils of both extant and extinct falcon species and subspecies from the early to late Pleistocene. (Data from Appendix)

2014; Núñez-Lahuerta et al. 2018). As previously noted, presumed subspecies and ancestors became extinct throughout the Early and Middle Pleistocene (see Sect. 1.4 and Fig. 1.9). Nevertheless, all 11 Early Pleistocene species in the genus *Falco* have survived to the present day (see also Sect. 1.4; Fig. 1.10; Finlayson et al. 2012b).

Overall, species with wide ecological adaptability are the most frequently encountered in fossil assemblages across all time periods (Fig. 1.10). Among large falcons, the peregrine falcon is the most common, followed by gyrfalcon and the saker falcon (see also the "Preface"). However, the presence of putative *F. antiquus* in the Early and Middle Pleistocene may mask the true frequency of saker falcons (Fig. 1.10; Mlikovský 2002). Lanner falcon fossils, which date back to the Middle Pleistocene, are relatively scarce, with only a few records (Leonardi 2015).

Predictably, the common kestrel is the most dominant falcon across all archaeological sites and time periods (Fig. 1.10). During the Early and Middle Pleistocene, it shared territories with the *F. t. atavus* subspecies but became the sole species present in the Late Pleistocene. Other small falcon species had similar frequencies of occurrence throughout the three Pleistocene periods (Fig. 1.10). Interestingly, these frequencies closely match the current distribution patterns. Consequently, with the exception of the gyrfalcon and red-footed falcon's brief expansion during the LGM, the historical distribution of falcons in the Western Palearctic is consistent with their current ranges (see Fig. 1.4 and species accounts).

The significant expansion of tundra habitat during the LGM favored Arctic and sub-Arctic falcon species (Arctic peregrine and gyrfalcon) as well as continental open-habitat species such as the saker falcon. As the tundra began to contract northward, peregrine falcon populations subsequently declined (Gu et al. 2021). During the Weichselian glacial period, both gyrfalcons and saker falcons coexisted in Europe as parapatric species. The separation of these populations occurred during the last interglacial period when they became isolated by the taiga belt (Fig. 1.11; Leonardi 2020).

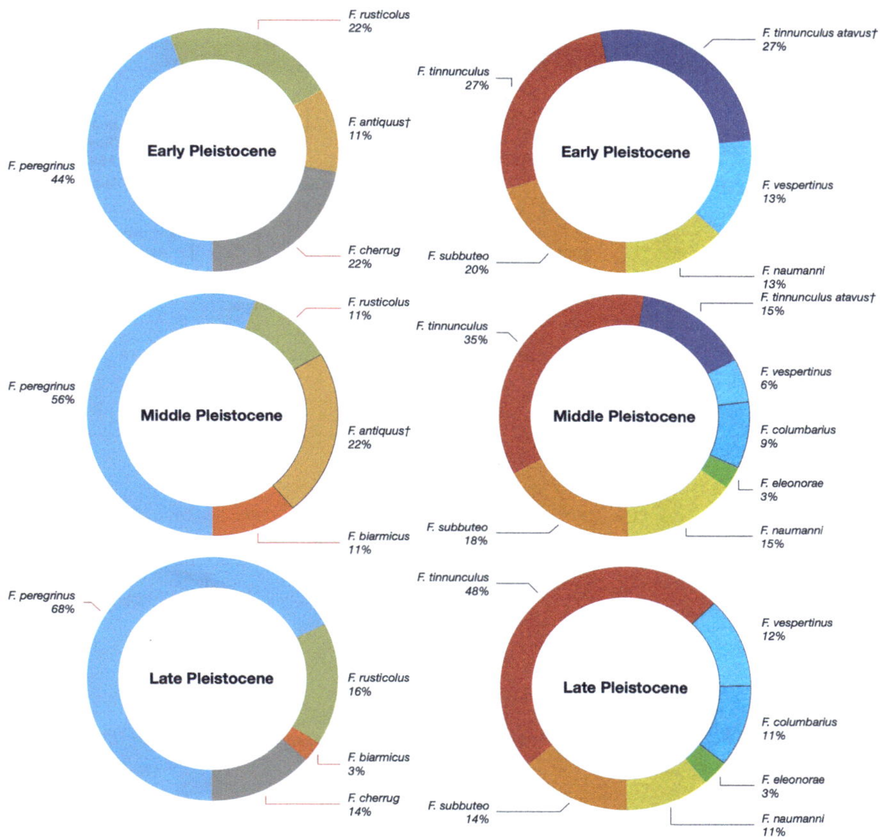

Fig. 1.10 Frequencies of occurrence of large (left) and small (right) falcon species across the Early, Middle, and Late Pleistocene periods (data from Appendix). Number of archaeological sites: Early Pleistocene = 14; Middle Pleistocene = 24; Late Pleistocene = 94

1.6.1 Falco naumanni

Throughout the Pleistocene, the lesser kestrel may have been present in parts of the Western Palearctic ecoregion, possibly utilizing refuges in Iberia during glacial periods (Figs. 1.10 and 1.12; Negro et al. 2020). Similar to other warm-adapted falcon species, it occupied the Mediterranean basin much as it does today (Figs. 1.4 and 1.12).

As cave-dwelling species, lesser kestrels shared breeding, roosting, and foraging habitats with early humans. Conversely, humans hunted lesser kestrels for food and valued their feathers for ornamental purposes (Peresani et al. 2011; Finlayson et al. 2012a, b; Negro et al. 2020). Except for Cova Negra, evidence of de-fleshing activities likely associated with raptor bones for food is rare and limited to MIS 3 (Rufà

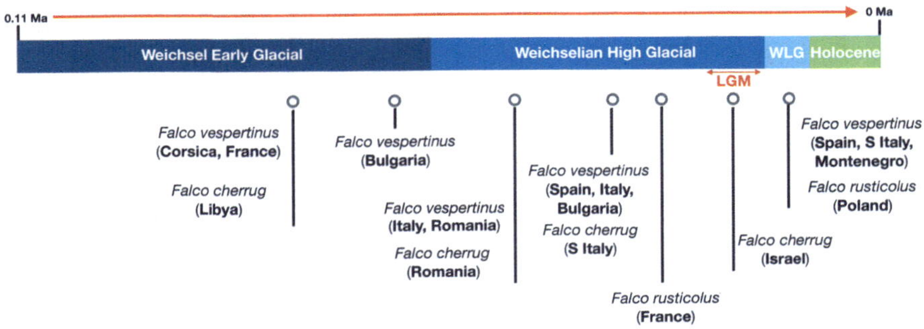

Fig. 1.11 Fossils of the genus *Falco* from glaciation periods (data from Appendix; WLG = Weichselian Late Glacial)

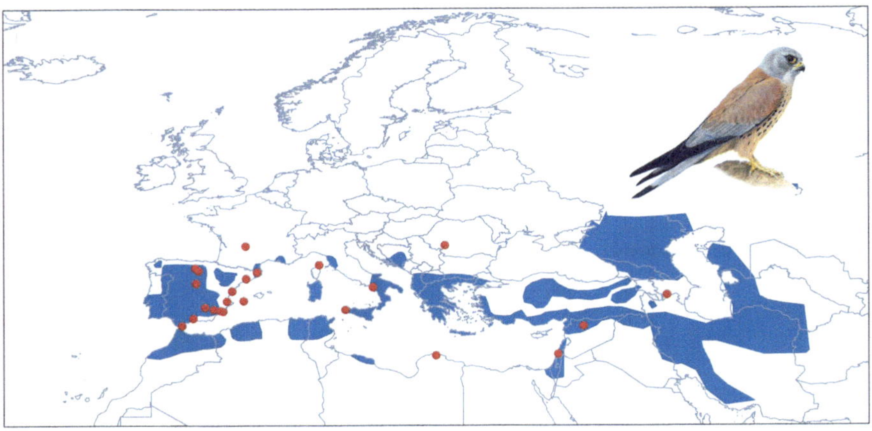

Fig. 1.12 Breeding range of the lesser kestrel in the Western Palearctic. Red dots indicate archaeological sites with fossilized remains of this species dating from the Early to Late Pleistocene. (Drawings courtesy of Marco Preziosi)

and Laroulandie 2021). Nonetheless, a lesser kestrel scapula from this site shows striae on its shaft (Martínez Valle et al. 2016).

1.6.2 Falco tinnunculus

Undoubtedly, the common kestrel is the most frequently encountered species of the genus *Falco* at archaeological sites and throughout the Pleistocene epoch (Fig. 1.10). The abundance of kestrel long bones, including complete skeletons of subadult birds, suggests that they bred in caves and that their offspring perished from natural

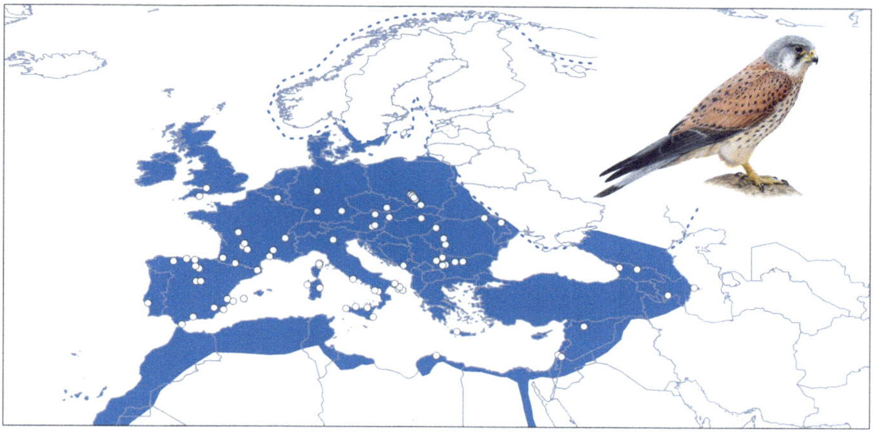

Fig. 1.13 The current breeding range of the common kestrel in the Western Palearctic. Boundaries of migrant breeding areas are outlined by a blue dashed line. White dots indicate archaeological sites with fossilized remains of this species from the Early to Late Pleistocene. (Drawings courtesy of Marco Preziosi)

causes (Gál 2008). Archaeological findings trace its current breeding range in the Western Palearctic, and its present distribution as a migrant breeder corresponds in part formerly ice-covered area (Fig. 1.13). Despite this, it was found in south-west England during earlier cold periods and the interglacial period (Harrison 1980). It did not reach the Canary Islands due to fluctuations in sea level, ocean currents, trade winds, and temperature, which repeatedly altered habitats (Kangas et al. 2018). Kestrels are responsible for bone accumulation of small mammals, birds, and lizards due to their predatory activities (Gál 2008; Guerra et al. 2013).

1.6.3 Falco vespertinus

The current distribution range of the red-footed falcon in the Western Palearctic extends across Central and Eastern Europe (Fig. 1.14). While entire populations migrate to Middle East and Southern Europe, individual falcons are also known to wander throughout the Western Palearctic (Fig. 1.14).

During colder periods, the red-footed falcon expanded westward due to the prevalence of dry continental conditions, occupying steppe and forest–steppe lowlands with dense grass cover (Fig. 1.14; Gál et al. 2003; Finlayson et al. 2016). Evidence of exploitation by Neanderthals (see Fig. 1.8), indicated by small carnivore gnaw marks, is also found on bones of red-footed falcons (Peresani et al. 2011; Oros Sršen et al. 2014).

Fig. 1.14 The current breeding range of the red-footed falcon in the Western Palearctic. Boundaries of migratory passage areas are indicated by a blue dashed line. Red dots indicate archaeological sites with fossilized remains of this species from the Early to Late Pleistocene. (Drawings courtesy of Marco Preziosi)

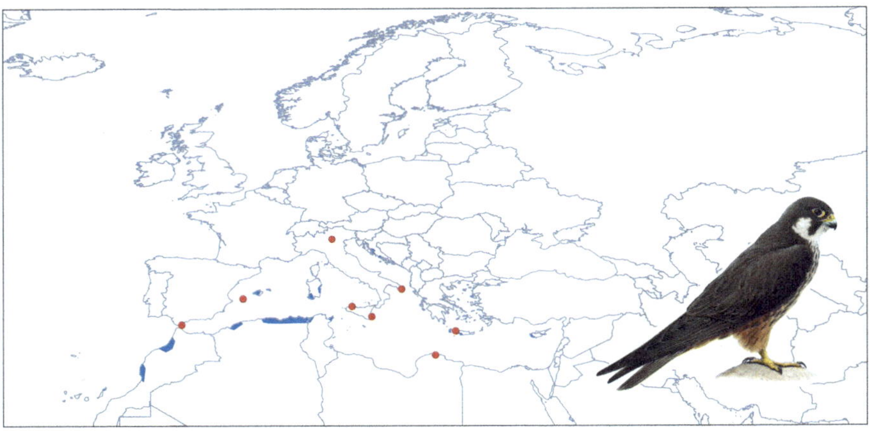

Fig. 1.15 The current breeding range of Eleonora's falcon in the Western Palearctic. Red dots indicate archaeological sites with fossilized remains of this species from the Early to Late Pleistocene. (Drawings courtesy of Marco Preziosi)

1.6.4 Falco eleonorae

Eleonora's falcon is a migrant breeding species native to the Mediterranean basin, where it nests primarily in colonies on small islands, islets, and inaccessible sea cliffs in the Mediterranean and north-western Atlantic Ocean (Fig. 1.15). Fossilized remains have been found at few archaeological sites in the southern Mediterranean, particularly on the nesting islands such as the Balearics, Crete, and Sicily (see Appendix; Fig. 1.15).

This falcon is a trans-Saharan migrant that winters mainly in Madagascar. As a Palearctic–Palaeotropical migratory species, its migration routes and patterns were likely established during the Pleistocene or earlier (Finlayson et al. 2012b). Interestingly, there are only two records of *Falconidae* fossils from the early Pliocene (2.66 Mya) in Africa. One of these, discovered at ancient sites near Laetoli in East Africa, is closely related to Eleonora's falcon (Louchart 2011).

1.6.5 Falco columbarius

Merlin has a global distribution, with a boreal breeding range and wintering areas in southern latitudes (Fig. 1.16). Like other primary genetic clades within the genus *Falco*, it is a migratory species that prefer open environments in temperate and boreal regions of the Northern Hemisphere (Bedetti and Pavia 2007; Fuchs et al. 2015; Leonardi 2020). The Bering Strait and the Atlantic Ocean likely contributed to significant divergence between Holarctic and Palearctic populations (Johnsen et al. 2010; Fuchs et al. 2015).

During the colder periods, the merlin experienced a significant shift southward in its preferred habitats, allowing it to occupy territories in the Mediterranean (Fig. 1.16). Its presence in fossil assemblages aligns with the boundaries of its non-breeding range (Fig. 1.16). Notably, the merlin is also a winter visitor to lowlands

Fig. 1.16 The current breeding range of the merlin in the Western Palearctic. Boundaries of native nonbreeding areas are denoted by a blue dashed line. Red dots indicate archaeological sites, with fossilized remains of this species from the Early to Late Pleistocene. (Drawings courtesy of Marco Preziosi)

in northern and central of Israel, as evidenced by fossils found in this country and Syria (Fig. 1.16).

1.6.6 Falco subbuteo

The Eurasian hobby has been present in the Western Palearctic since the Early Pleistocene but became widespread across Europe during the Middle Pleistocene (Tyrberg 2008). This species is well-adapted to dry open areas adjacent to woodlands, forest steppes, riverine forests, and forest edges and is found in both continental and southern Europe (Fig. 1.17). As a bioclimatic generalist with a strong migratory tendency, this species has an extensive breeding range that encompasses numerous archaeological sites where its fossilized remains have been discovered (Fig. 1.17).

1.6.7 Falco biarmicus

The lanner falcon's distribution in the Western Palearctic is fragmented, with numerous widespread population demes (Fig. 1.18). Fossil remains have been discovered only in Israel and Azerbaijan (Appendix; Tyrberg 2008a). Interestingly, there are no fossils from continental Europe (Fig. 1.18). After a comprehensive review, a large collection of bones discovered in Corsica, previously attributed to the lanner falcon,

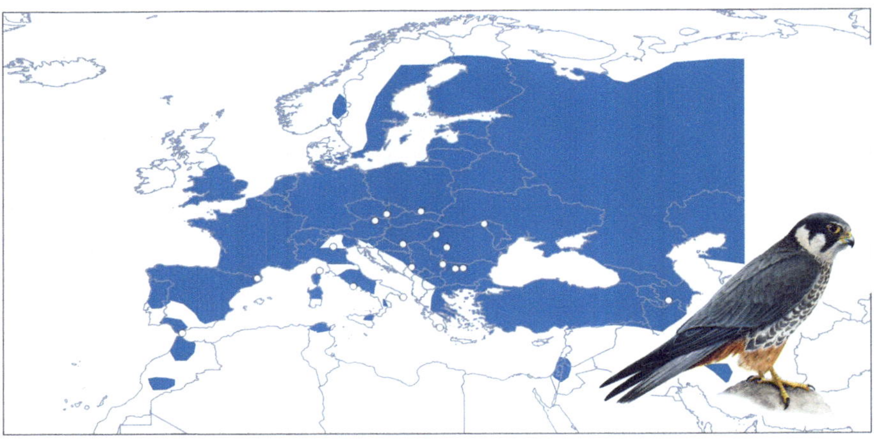

Fig. 1.17 The current breeding range of the Eurasian hobby in the Western Palearctic. White dots indicate archaeological sites with fossilized remains of this species from the Early to Late Pleistocene. (Drawings courtesy of Marco Preziosi)

Fig. 1.18 The current breeding range of the lanner falcon in the Western Palearctic. Red dots indicate archaeological sites with fossilized remains of this species from the Early to Late Pleistocene. (Drawings courtesy of Marco Preziosi)

was instead recognized as belonging to a new subspecies of the peregrine falcon, *F. p. cyrneus* (Louchart 2002).

Recent excavations at the archaeological site of Ohalo 2 near the Sea of Galilee in Israel uncovered two lanner falcons (Simmons and Nadel 1998). In these areas, birds of prey were frequently exploited by humans, primarily as a food source but also for the cultural and decorative values of their claws and feathers (Simmons and Nadel 1998).

1.6.8 Falco cherrug

Saker falcon fossilized remains are scarce in the Western Palearctic, partially due to dubious identifications, with several bones simply classified as "*large falcon*" (Bedetti and Pavia 2007; Tyrberg 2008a; Boev and Tsoukala 2019; see also Sect. 1.4 and Fig. 1.7). Some of these locations are slightly outside of the species' current breeding range (Fig. 1.19), while others are in present-day wintering areas, such as Israel and Apulia (southern Italy), which are the westernmost locations in the Western Palearctic (Fig. 1.16; Bedetti and Pavia 2007). The saker falcon is primarily found in open areas with scattered trees, riverine environment, open forests and plains, grass and tree steppes, and mountain plateaus. During the Weichselian glaciation, the saker falcon and the gyrfalcon coexisted in Europe as parapatric species and exhibit perfect synteny (Joseph et al. 2018).

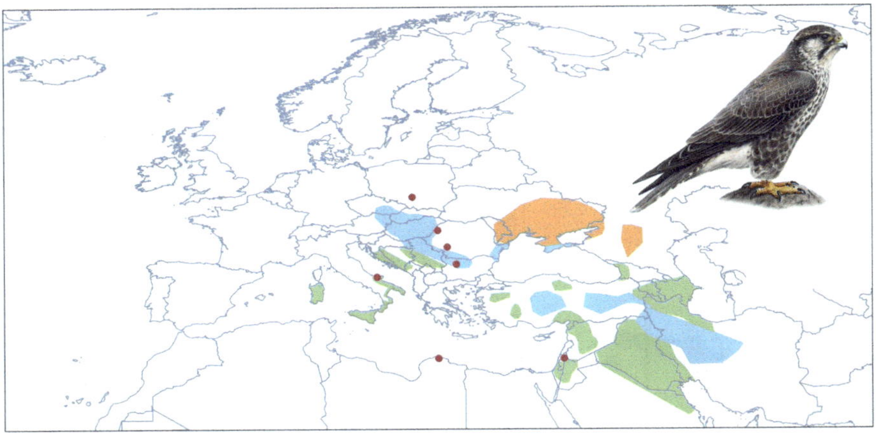

Fig. 1.19 The distributional range of the saker falcon in the Western Palearctic (light green = native nonbreeding; light blue = native resident; orange = native breeding; drawings courtesy of Marco Preziosi)

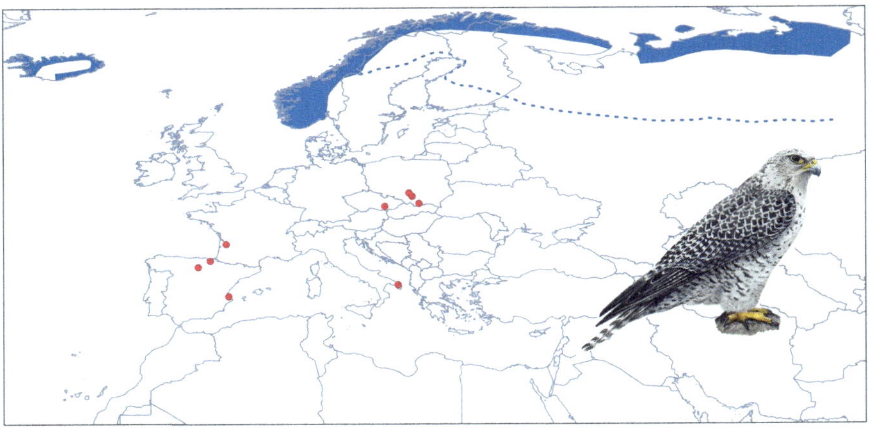

Fig. 1.20 The current breeding range of the gyrfalcon in the Western Palearctic. Boundaries of wintering areas are denoted by a blue dashed line. Red dots indicate archaeological sites with fossilized remains of this species from the Early to Late Pleistocene. (Drawings courtesy of Marco Preziosi)

1.6.9 Falco rusticolus

The gyrfalcon is an Arctic predator that expanded rapidly from a single population during the late Pleistocene, eventually occupying its current circumpolar breeding range (Johnson et al. 2007). Favourable cold climate conditions enabled its expansion into southern Europe, Asia, and western United States (Fig. 1.20; Johnson et al.

2007; Tyrberg 2008a). During the LGM, the gyrfalcon's main prey (willow and rock ptarmigans) also benefited from these climatic circumstances, which allowed them to expand their breeding ranges (Huntley and Green 2011). However, no relict gyrfalcon populations survived the glacial expansion (Tyrberg 1991).

Although gyrfalcons do not nest in caves, they are considered fossil accumulators because the remains of their avian and mammalian prey often fall into caves and accumulate there (Bocheński et al. 1998). Their use of the same nest sites over extended periods has also provided valuable insights into paleoenvironmental conditions (see Chap. 2; Burnham et al. 2009).

1.6.10 Falco peregrinus

The peregrine falcon is one of the few species worldwide which breeds across nearly seven ecozones (Newton and Dale 2001). As expected, it is commonly found in Pleistocene fossil assemblages (Fig. 1.10). The species likely underwent rapid expansion during the late Pleistocene and early Holocene (White et al. 2013). Fossil remains from this period have been discovered outside the Western Palearctic, in locations ranging from the United States to Fiji to the Egyptian–Sudanese border (Tchernov 1968; Porter and White 1973; Talbot et al. 2011).

Primarily inhabiting open environments, including Arctic tundra, the peregrine falcon is less common in forested areas. The current distribution of native resident populations corresponds with the locations of fossil remains (Fig. 1.21). Many peregrine falcon populations from northern and continental regions migrate through areas that were covered by perennial ice during the Pleistocene (Fig. 1.21).

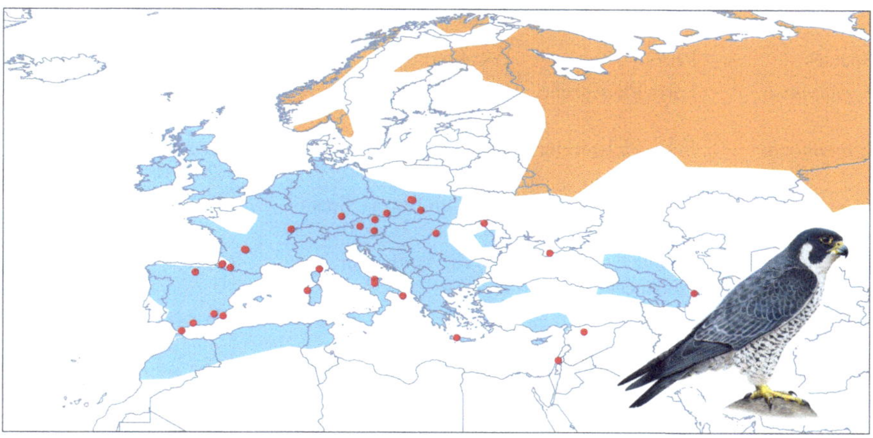

Fig. 1.21 The distributional range of the peregrine falcon in the Western Palearctic (light blue = native resident; orange = native breeding). (Drawings courtesy of Marco Preziosi)

1.7 Conclusions

During the early Pleistocene, the number of falcon species in the Western Palearctic was limited, but this increased through the middle Pleistocene and became more consolidated by the late Pleistocene (see Appendix). There is a notable disparity between the number of archaeological sites from different periods, with a clear predominance for more recent Pleistocene sites. Despite this, the composition percentages in fossil assemblages have shown some stability. Ultimately, all falcon species extant during the Early Pleistocene have survived to the present day.

Several key points can be highlighted:

- Past distribution areas generally align with current ones.
- Presumed subspecies and ancestors became extinct during the Early and Middle Pleistocene.
- significant latitudinal shifts (e.g., gyrfalcon, merlin, and Arctic peregrine) and longitudinal shifts (e.g., red-footed falcon) occurred during the LGM.
- No relict populations survived from the glacial range expansion.
- Warm-adapted species (e.g., lesser kestrels and Eleonora's falcons) continued to inhabit the Mediterranean basin as they do today.
- The scarcity of fossil remains for certain species, such as the lanner falcon, limits to fully understand their distribution patterns during the Pleistocene.

Appendix: Fossil Remains of Western Palearctic Falcons from the Pleistocene

Species	Period	Location	References
F. naumanni	Early Pleistocene	Cueva Victoria (Murcia), Spain	Tyrberg (2008a)
F. naumanni	Early Pleistocene	Gran Dolina, Spain	Cuenca-Bescós et al. (2013)
F. naumanni	Middle Pleistocene	Cau d'en Borrás, Spain	Tyrberg (2008a)
F. naumanni	Middle Pleistocene	Azykh, Azerbaijan	Tyrberg (2008a)
F. naumanni	Middle Pleistocene	Sima Palomas Cabezo Gordo, Spain	Tyrberg (2008a)
F. naumanni	Middle Pleistocene	Pinilla del Valle, Spain	Tyrberg (2008a)
F. naumanni	Middle Pleistocene	Abîmes de La Fage, France	Mourer-Chauviré et al. (2003)
F. naumanni	Late Pleistocene	Cova Negra, Spain	Martínez Valle et al. (2016)

(continued)

Species	Period	Location	References
F. naumanni	Late Pleistocene	Gorham's Cave, Gibraltar	Finlayson et al. (2016)
F. naumanni	Late Pleistocene	Zafarraya, Spain	Finlayson et al. (2016)
F. naumanni	Late Pleistocene	Grotte de la Coscia (Corse), France	Tyrberg (2008a)
F. naumanni	Late Pleistocene	Hayonim cave (Galilee), Israel	Tyrberg (2008a)
F. naumanni	Late Pleistocene	Arbreda Cave, Spain	Lloveras et al. (2018)
F. naumanni	Late Pleistocene	Boroşten, Romania	Tyrberg (2008a)
F. naumanni	Late Pleistocene	Cueva de Ambrosio, Spain	Tyrberg (2008a)
F. naumanni	Late Pleistocene	Cueva de Valdegoba, Spain	Tyrberg (2008a)
F. naumanni	Late Pleistocene	Es Pouàs (Eivissa), Spain	Tyrberg (2008a)
F. naumanni	Late Pleistocene	Castelcivita, Italy	Tyrberg (2008a)
F. naumanni	Late Pleistocene	Tell Mureybet, Syria	Tyrberg (2008a)
F. naumanni	Late Pleistocene	Cueva Negra de la Incarnación, Spain	Tyrberg (2008a)
F. naumanni	Late Pleistocene	Grotta del Faraglione, Italy	Tyrberg (2008a)
F. naumanni	Late Pleistocene	Haua Fteah, Libya	Tyrberg (2008a)
F. naumanni	Late Pleistocene	Cova de la Guineu	Tyrberg (2008a)
F. naumanni	Late Pleistocene	Pech-de-l'Azé, France	Eastham (2021)
F. tinnunculus	Early Pleistocene	Sima del elefante, Spain	Núñez-Lahuerta et al. (2021)
F. tinnunculus	Early Pleistocene	Slivnitsa, Bulgaria	Tyrberg (2008a)
F. tinnunculus	Early Pleistocene	Betfia, Romania	Tyrberg (2008a)
F. tinnunculus	Early Pleistocene	Pinilla del Valle, Spain	Tyrberg (2008a)
F. tinnunculus	Middle Pleistocene	Azykh, Azerbaijan	Tyrberg (2008a)
F. tinnunculus	Middle Pleistocene	Grotte Vaufrey, France	Rufà and Laroulandie (2021)
F. tinnunculus	Middle Pleistocene	Stránská Skála, Czech Republic	Tyrberg (2008a)
F. tinnunculus	Middle Pleistocene	Grotta di Spinagallo, Italy	Tyrberg (2008a)
F. tinnunculus	Middle Pleistocene	Grotta del Ponte, Italy	Tyrberg (2008a)
F. tinnunculus	Middle Pleistocene	Sima Palomas Cabezo Gordo, Spain	Tyrberg (2008a)
F. tinnunculus	Middle Pleistocene	Biśnik Cave, Poland	Tomek et al. (2012)
F. tinnunculus	Middle Pleistocene	Hundsheim (Niederösterreich), Austria	Mlíkovský (2009)
F. tinnunculus	Middle/Late Pleistocene	Binagady, Azerbaijan	Tyrberg (2008a)
F. tinnunculus	Middle/Late Pleistocene	Kudaro 1 (South Ossetia), Georgia	Tyrberg (2008a)
F. tinnunculus	Middle/Late Pleistocene	Marzal II (Ardèche), France	Tyrberg (2008a)

(continued)

Species	Period	Location	References
F. tinnunculus	Middle/Late Pleistocene	Tour-de-Faire, France	Tyrberg (2008a)
F. tinnunculus	Late Pleistocene	Abrigo do Lagar Velho, Portugal	Moreno-Garcia and Pimenta (2002)
F. tinnunculus	Late Pleistocene	Furninha Cave, Portugal	Figueiredo et al. (2022)
F. tinnunculus	Late Pleistocene	Grotte de la Coscia (Corse), France	Tyrberg (2008a)
F. tinnunculus	Late Pleistocene	Cova Negra, Spain	Martínez Valle et al. (2016)
F. tinnunculus	Late Pleistocene	Grosse Ofenbergerhöhle, Austria	Tyrberg (2008a)
F. tinnunculus	Late Pleistocene	Chelm's Combe Shelter, UK	Stewart (2007)
F. tinnunculus	Late Pleistocene	Gudenushöhle, Austria	Tyrberg (2008a)
F. tinnunculus	Late Pleistocene	Luegloch, Austria	Tyrberg (2008a)
F. tinnunculus	Late Pleistocene	Trou de Chaleux (Namur), Belgium	Tyrberg (2008a)
F. tinnunculus	Late Pleistocene	Gorham's Cave, Gibraltar	Finlayson et al. (2016)
F. tinnunculus	Late Pleistocene	Zafarraya, Spain	Finlayson et al. (2016)
F. tinnunculus	Late Pleistocene	Grotta di Fumane, Italy	Peresiani et al. (2011)
F. tinnunculus	Late Pleistocene	Koziarnia, Poland	Berto et al. (2021)
F. tinnunculus	Late Pleistocene	Cave no. 16, Bulgaria	Tyrberg (2008a)
F. tinnunculus	Late Pleistocene	Temnata Cave, Bulgaria	Tyrberg (2008a)
F. tinnunculus	Late Pleistocene	Abri Dufaure, France	Tyrberg (2008a)
F. tinnunculus	Late Pleistocene	Bois-Ragot, France	Tyrberg (2008a)
F. tinnunculus	Late Pleistocene	Grotta di Funtanedu, Corse, France	Tyrberg (2008a)
F. tinnunculus	Late Pleistocene	Grotte Jean-Pierre, France	Tyrberg (2008a)
F. tinnunculus	Late Pleistocene	Combe Saunière 1, France	Tyrberg (2008a)
F. tinnunculus	Late Pleistocene	Grotta del Cavallo, Italy	Carrera et al. (2021)
F. tinnunculus	Late Pleistocene	Pešturina, Serbia	Mihailovíc et al. (2022)
F. tinnunculus	Late Pleistocene	Okumi I (Abkhazia), Georgia	Tyrberg (2008a)
F. tinnunculus	Late Pleistocene	Liko cave (Crete), Greece	Tyrberg (2008a)
F. tinnunculus	Late Pleistocene	Vaskapu (Bükk), Hungary	Tyrberg (2008a)
F. tinnunculus	Late Pleistocene	Hayonim cave (Galilee), Israel	Tyrberg (2008a)
F. tinnunculus	Late Pleistocene	Cala della Calcina (Sardegna), Italy	Tyrberg (2008a)
F. tinnunculus	Late Pleistocene	Grotta Rapaci (Sardegna), Italy	Tyrberg (2008a)
F. tinnunculus	Late Pleistocene	Grotta Romanelli (Puglia), Italy	Tyrberg (2008a)
F. tinnunculus	Late Pleistocene	Arbreda Cave, Spain	Lloveras et al. (2018)
F. tinnunculus	Late Pleistocene	Crvena stijena, Montenegro	Dogandžić and Đuričić (2017)
F. tinnunculus	Late Pleistocene	Grotta della Madonna, Italy	Tyrberg (2008a)
F. tinnunculus	Late Pleistocene	Grotta di Cardamone, Italy	Tyrberg (2008a)

(continued)

Species	Period	Location	References
F. tinnunculus	Late Pleistocene	Brinzeni, Moldova	Tyrberg (2008a)
F. tinnunculus	Late Pleistocene	Gruta da Figueira Brava, Portugal	Tyrberg (2008a)
F. tinnunculus	Late Pleistocene	Boroşten, Romania	Tyrberg (2008a)
F. tinnunculus	Late Pleistocene	Nandru, Romania	Tyrberg (2008a)
F. tinnunculus	Late Pleistocene	Cueva de Cendres, Spain	Tyrberg (2008a)
F. tinnunculus	Late Pleistocene	Cueto de la Mina, Spain	Tyrberg (2008a)
F. tinnunculus	Late Pleistocene	Cueva de Valdegoba, Spain	Tyrberg (2008a)
F. tinnunculus	Late Pleistocene	Cueva del Castillo, Spain	Tyrberg (2008a)
F. tinnunculus	Late Pleistocene	Es Pouàs (Eivissa), Spain	Tyrberg (2008a)
F. tinnunculus	Late Pleistocene	Windmill Hill Cave, Brixham, UK	Harrison (1980)
F. tinnunculus	Late Pleistocene	Komarowa Cave, Poland	Tyrberg (2008a)
F. tinnunculus	Late Pleistocene	Grotta di San Teodoro, Italy	Tyrberg (2008a)
F. tinnunculus	Late Pleistocene	Vykhvatintsy cave, Moldova	Tyrberg (2008a)
F. tinnunculus	Late Pleistocene	Oblazowa Cave, Poland	Tyrberg (2008a)
F. tinnunculus	Late Pleistocene	Krucza Skala, Poland	Tyrberg (2008a)
F. tinnunculus	Late Pleistocene	Tell Mureybet, Syria	Tyrberg (2008a)
F. tinnunculus	Late Pleistocene	Cueva Negra de la Incarnación, Spain	Tyrberg (2008a)
F. tinnunculus	Late Pleistocene	Kozarnika cave, Bulgaria	Tyrberg (2008a)
F. tinnunculus	Late Pleistocene	Grotta del Faraglione, Italy	Tyrberg (2008a)
F. tinnunculus	Late Pleistocene	San Maria d'Agnano, Italy	Tyrberg (2008a)
F. tinnunculus	Late Pleistocene	Ohalo II, Israel	Tyrberg (2008a)
F. tinnunculus	Late Pleistocene	Gritulu Cave, France	Tyrberg (2008a)
F. tinnunculus	Late Pleistocene	Deszczowa Cave, Poland	Tyrberg (2008a)
F. tinnunculus	Late Pleistocene	Sesselfelsgrotte, Germany	Tyrberg (2008a)
F. tinnunculus	Late Pleistocene	La Riera, Spain	Tyrberg (2008a)
F. tinnunculus	Late Pleistocene	Devetashka Cave, Bulgaria	Tyrberg (2008a)
F. tinnunculus	Late Pleistocene	Filipovska Cave, Bulgaria	Tyrberg (2008a)
F. tinnunculus	Late Pleistocene	Haua Fteah, Libya	Tyrberg (2008a)
F. tinnunculus	Late Pleistocene	Grotte de Noisetier, France	Tyrberg (2008a)
F. tinnunculus	Late Pleistocene	Jarama 2, Spain	Tyrberg (2008a)
F. tinnunculus	Late Pleistocene	Grotta della Serratura, Italy	Tyrberg (2008a)
F. tinnunculus	Late Pleistocene	Buhlen (Hessen), Germany	Tyrberg (2008a)
F. tinnunculus	Late Pleistocene	Grotta Breuil, Italy	Tyrberg (2008a)
F. tinnunculus	Late Pleistocene	Prepoštská Cave, Slovakia	Šedivá et al. (2020)
F. tinnunculus	Late Pleistocene	Buso Doppio del Broion, Italy	Carrera (2018)
F. tinnunculus	Late Pleistocene	Šandalja II, Croatia	Oros Sršen et al. (2014)
F. tinnunculus	Late Pleistocene	Grotta di Uluzzo, Italy	Milliken (1991)
F. tinnunculus	Late Pleistocene	Le Flageolet, France	Depelch (1983)
F. tinnunculus	Late Pleistocene	Pontnewydd cave, Clwyd, UK	Eastham (2021)

(continued)

Species	Period	Location	References
F. tinnunculus	Late Pleistocene	Pinhole Cave, UK	Eastham (2021)
F. tinnunculus	Late Pleistocene	La Cotte de St Brelade, Jersey	Eastham (2021)
F. tinnunculus	Late Pleistocene	Baume de Gigny, France	Eastham (2021)
F. tinnunculus	Late Pleistocene	Pech-de-l'Azé, France	Eastham (2021)
F. tinnunculus	Late Pleistocene	Hortus Herault, France	Eastham (2021)
F. tinnunculus	Late Pleistocene	Salpêtre, Herault; France	Eastham (2021)
F. tinnunculus	Late Pleistocene	Soulabé, Ariège, France	Eastham (2021)
F. tinnunculus	Late Pleistocene	Abri Olha. France	Eastham (2021)
F. tinnunculus	Late Pleistocene	Torre Nave, Italy	Eastham (2021)
F. tinnunculus	Late Pleistocene	Grotta del Principe, Italy	Eastham (2021)
F. tinnunculus	Late Pleistocene	InstÁllósko, Hungary	Eastham (2021)
F. tinnunculus	Late Pleistocene	Flageolet, France	Eastham (2021)
F. tinnunculus	Late Pleistocene	La Ferrassie, France	Eastham (2021)
F. tinnunculus	Late Pleistocene	Arene Candide, Italy	Eastham (2021)
F. tinnunculus atavus†	Early Pleistocene	Betfia, Romania	Tyrberg (2008a)
F. tinnunculus atavus†	Early Pleistocene	Beremend, Hungary	Tyrberg (2008a)
F. tinnunculus atavus†	Early Pleistocene	Deutsch-Altenburg, Austria	Tyrberg (2008a)
F. tinnunculus atavus†	Early Pleistocene	Somssichhégy 2 (Villány), Hungary	Tyrberg (2008a)
F. tinnunculus atavus†	Middle Pleistocene	Abîmes de La Fage, France	Mourer-Chauviré et al. (2003)
F. tinnunculus atavus†	Middle Pleistocene	Prezletice, Czech Republic	Tyrberg (2008a)
F. tinnunculus atavus†	Middle Pleistocene	Stránská Skála, Czech Republic	Tyrberg (2008a)
F. tinnunculus atavus†	Middle Pleistocene	Orgnac 3, France	Mourer-Chauviré (1975)
F. tinnunculus atavus†	Middle Pleistocene	Kozi Grzbiet, Poland	Bocheński (1984)
F. tinnunculus atavus†	Stratigraphy unkonwn	Biśnik Cave, Poland	Tomek et al. (2012)
F. vespertinus	Early Pleistocene	Somssichhégy 2 (Villány), Hungary	Tyrberg (2008a)
F. vespertinus	Early Pleistocene	Betfia, Romania	Tyrberg (2008a)
F. vespertinus	Middle Pleistocene	Azykh, Azerbaijan	Tyrberg (2008a)
F. vespertinus	Middle/Late Pleistocene	Stránská Skála, Czech Republic	Tyrberg (2008a)
F. vespertinus?	Late Pleistocene	Grotta di Fumane, Italy	Peresiani et al. (2011)
F. vespertinus	Late Pleistocene	Gorham's Cave, Gibraltar	Finlayson et al. (2016)
F. vespertinus	Late Pleistocene	Cave no. 16, Bulgaria	Tyrberg (2008a)
F. vespertinus	Late Pleistocene	Temnata Cave, Bulgaria	Tyrberg (2008a)

(continued)

Species	Period	Location	References
F. vespertinus	Late Pleistocene	Grotte de la Coscia (Corse), France	Tyrberg (2008a)
F. vespertinus	Late Pleistocene	Arbreda Cave, Spain	Lloveras et al. (2018)
F. vespertinus	Late Pleistocene	Kalamakia cave, Greece	Harvati et al. (2013)
F. vespertinus	Late Pleistocene	Grotta del Fossellone, Italy	Tyrberg (2008a)
F. vespertinus	Late Pleistocene	Brinzeni, Moldova	Tyrberg (2008a)
F. vespertinus	Late Pleistocene	Malisina stijena, Montenegro	Tyrberg (2008a)
F. vespertinus	Late Pleistocene	Pestera Bordu Mare, Romania	Tyrberg (2008a)
F. vespertinus	Late Pleistocene	Oblazowa Cave, Poland	Tyrberg (2008a)
F. vespertinus	Late Pleistocene	Kozarnika cave, Bulgaria	Tyrberg (2008a)
F. vespertinus	Late Pleistocene	Devetashka Cave, Bulgaria	Tyrberg (2008a)
F. vespertinus	Late Pleistocene	Teufelsrast-Knochenfuge, Austria	Tyrberg (2008a)
F. vespertinus	Late Pleistocene	Buso Doppio del Broion, Italy	Carrera (2018)
F. vespertinus	Late Pleistocene	Šandalja II, Croatia	Oros Sršen et al. (2014)
F. vespertinus	Late Pleistocene	Baume de Gigny, France	Eastham (2021)
F. vespertinus	Late Pleistocene	Soulabé, Ariège, France	Eastham (2021)
F. vespertinus	Late Pleistocene	Torre Nave, Italy	Eastham (2021)
F. vespertinus?	Late Pleistocene	Grotta del Principe, Italy	Eastham (2021)
F. vespertinus	Late Pleistocene	InstÁllósko, Hungary	Eastham (2021)
F. vespertinus	Late Pleistocene	Arene Candide, Italy	Eastham (2021)
F. eleonorae	Middle Pleistocene	Grotta di Poggio Schinaldi, Italy	Tyrberg (2008a)
F. eleonorae	Late Pleistocene	Gorham's Cave, Gibraltar	Finlayson et al. (2016)
F. eleonorae	Late Pleistocene	Liko cave (Crete), Greece	Tyrberg (2008a)
F. eleonorae	Late Pleistocene	Grotta Romanelli (Puglia), Italy	Tyrberg (2008a)
F. eleonorae	Late Pleistocene	Es Pouàs (Eivissa), Spain	Tyrberg (2008a)
F. eleonorae	Late Pleistocene	Haua Fteah, Libya	Tyrberg (2008a)
F. eleonorae	Late Pleistocene	Arene Candide, Italy	Eastham (2021)
F. columbarius	Middle Pleistocene	Grotta di Spinagallo, Italy	Tyrberg (2008a)
F. columbarius	Middle Pleistocene	Contrada Fusco, Italy	Tyrberg (2008a)
F. columbarius	Middle Pleistocene	Abîmes de La Fage, France	Mourer-Chauviré et al. (2003)
F. columbarius	Late Pleistocene	Koziarnia, Poland	Berto et al. (2021)
F. columbarius	Late Pleistocene	Untertürkheim, Germany	Tyrberg (2008a)
F. columbarius	Late Pleistocene	Grotte de la Coscia (Corse), France	Tyrberg (2008a)
F. columbarius	Late Pleistocene	L'Esquicho-Grapaou (Gard), France	Tyrberg (2008a)
F. columbarius	Late Pleistocene	Kudaro 1 (South Ossetia), Georgia	Tyrberg (2008a)

(continued)

Species	Period	Location	References
F. columbarius	Late Pleistocene	Ingarano, Italy	Bedetti and Pavia (2007)
F. columbarius	Late Pleistocene	Malisina stijena, Montenegro	Tyrberg (2008a)
F. columbarius	Late Pleistocene	Boroşten, Romania	Tyrberg (2008a)
F. columbarius	Late Pleistocene	Komarowa Cave, Poland	Tyrberg (2008a)
F. columbarius	Late Pleistocene	Grotta di San Teodoro, Italy	Tyrberg (2008a)
F. columbarius	Late Pleistocene	Tell Mureybet, Syria	Tyrberg (2008a)
F. columbarius	Late Pleistocene	San Maria d'Agnano, Italy	Tyrberg (2008a)
F. columbarius	Late Pleistocene	Ohalo II, Israel	Tyrberg (2008a)
F. columbarius	Late Pleistocene	Sesselfelsgrotte, Germany	Tyrberg (2008a)
F. columbarius	Late Pleistocene	Grotta del Cavallo, Italy	Tyrberg (2008a)
F. columbarius	Late Pleistocene	Prepoštská Cave, Slovakia	Šedivá et al. (2020)
F. columbarius	Late Pleistocene	Grotta di Fumane, Italy	Romandini et al. (2016)
F. columbarius	Late Pleistocene	Salpêtre, Herault, France	Eastham (2021)
F. columbarius	Late Pleistocene	Flageolet, France	Eastham (2021)
F. subbuteo	Early Pleistocene	Cave Pirro-dell'Erba, Italy	Tyrberg (2008a)
F. subbuteo	Early Pleistocene	Pirro Nord, Italy	Bedetti and Pavia (2013)
F. subbuteo	Early Pleistocene	Betfia, Romania	Tyrberg (2008a)
F. subbuteo	Middle Pleistocene	Stránská Skála, Czech Republic	Tyrberg (2008a)
F. subbuteo	Middle Pleistocene	Kudaro 1 (South Ossetia), Georgia	Tyrberg (2008a)
F. subbuteo	Middle Pleistocene	Grotta di Spinagallo, Italy	Tyrberg (2008a)
F. subbuteo	Middle Pleistocene	Grotta di Marasà, Italy	Tyrberg (2008a)
F. subbuteo	Middle Pleistocene	Abîmes de La Fage, France	Mourer-Chauviré et al. (2003)
F. subbuteo	Middle/Late Pleistocene	Tropfsteinhöhle, Austria	Tyrberg (2008a)
F. subbuteo	Late Pleistocene	Temnata Cave, Bulgaria	Tyrberg (2008a)
F. subbuteo	Late Pleistocene	Grotte Jean-Pierre, France	Tyrberg (2008a)
F. subbuteo	Late Pleistocene	Grotte de la Coscia (Corse), France	Tyrberg (2008a)
F. subbuteo	Late Pleistocene	Liko cave (Crete), Greece	Tyrberg (2008a)
F. subbuteo	Late Pleistocene	Hayonim cave (Galilee), Israel	Tyrberg (2008a)
F. subbuteo	Late Pleistocene	Grotta Romanelli (Puglia), Italy	Tyrberg (2008a)
F. subbuteo	Late Pleistocene	Arbreda Cave, Spain	Lloveras et al. (2018)
F. subbuteo	Late Pleistocene	Grotta della Madonna, Italy	Tyrberg (2008a)
F. subbuteo	Late Pleistocene	Grotta di Cardamone, Italy	Tyrberg (2008a)
F. subbuteo	Late Pleistocene	Malisina stijena, Montenegro	Tyrberg (2008a)
F. subbuteo	Late Pleistocene	Komarowa Cave, Poland	Tyrberg (2008a)
F. subbuteo	Late Pleistocene	Castelcivita, Italy	Tyrberg (2008a)
F. subbuteo	Late Pleistocene	Oblazowa Cave, Poland	Tyrberg (2008a)

(continued)

Species	Period	Location	References
F. subbuteo	Late Pleistocene	Tell Mureybet, Syria	Tyrberg (2008a)
F. subbuteo	Late Pleistocene	Tsareva tsarkva, Bulgaria	Tyrberg (2008a)
F. subbuteo	Late Pleistocene	San Maria d'Agnano, Italy	Tyrberg (2008a)
F. subbuteo	Late Pleistocene	Lukenjska jama, Slovenia	Tyrberg (2008a)
F. subbuteo	Late Pleistocene	Sesselfelsgrotte, Germany	Tyrberg (2008a)
F. subbuteo	Late Pleistocene	Devetashka Cave, Bulgaria	Tyrberg (2008a)
F. subbuteo	Late Pleistocene	Bacon Hole, UK	Tyrberg (2008a)
F. subbuteo	Late Pleistocene	Baume de Gigny, France	Eastham (2021)
F. subbuteo	Late Pleistocene	Pech-de-l'Azé, France	Eastham (2021)
F. subbuteo	Late Pleistocene	Šandalja II, Croatia	Eastham (2021)
F. subbuteo	Late Pleistocene	Grotta di Fumane, Italy	Eastham (2021)
F. subbuteo	Late Pleistocene	Vindija, Croatia	Eastham (2021)
F. biarmicus	Middle Pleistocene	Azykh, Azerbaijan	Tyrberg (2008a)
F. biarmicus	Late Pleistocene	Ohalo II, Israel	Tyrberg (2008a)
F. cherrug	Early Pleistocene	Betfia, Romania	Tyrberg (2008a)
F. cherrug	Early Pleistocene	Kunino, Bulgaria	Tyrberg (2008a)
F. cherrug	Late Pleistocene	Ingarano, Italy	Bedetti and Pavia (2007)
F. cherrug	Late Pleistocene	Boroşten, Romania	Tyrberg (2008a)
F. cherrug	Late Pleistocene	Komarowa Cave, Poland	Tyrberg (2008a)
F. cherrug	Late Pleistocene	Ohalo II, Israel	Tyrberg (2008a)
F. cherrug	Late Pleistocene	Haua Fteah, Libya	Tyrberg (2008a)
F. rusticolus	Early Pleistocene	Sima del elefante, Spain	Núñez-Lahuerta et al. (2021)
F. rusticolus	Early Pleistocene	Stránská Skála, Czech Republic	Mlíkovský (2002)
F. rusticolus	Middle Pleistocene	Biśnik Cave, Poland	Tomek et al. (2012)
F. rusticolus	Late Pleistocene	Cueva del Volcán del Faro, Spain	Davidson (1989)
F. rusticolus	Late Pleistocene	Urtiaga, Spain	Elorza (1990)
F. rusticolus	Late Pleistocene	Pair-non-Pair (Gironde), France	Tyrberg (2008a)
F. rusticolus?	Late Pleistocene	Grotta di Cardamone, Italy	Tyrberg (2008a)
F. rusticolus	Late Pleistocene	Komarowa Cave, Poland	Tyrberg (2008a)
F. rusticolus	Late Pleistocene	Oblazowa Cave, Poland	Tyrberg (2008a)
F. rusticolus?	Late Pleistocene	Vindija, Croatia	Eastham (2021)
F. peregrinus	Early Pleistocene	Cave Pirro-dell'Erba, Italy	Tyrberg (2008a)
F. peregrinus	Early Pleistocene	Pirro Nord, Italy	Bedetti and Pavia (2013)
F. peregrinus	Early Pleistocene	Betfia, Romania	Tyrberg (2008a)
F. peregrinus	Early Pleistocene	Cueva Victoria, Murcia, Spain	Tyrberg (2008a)
F. peregrinus	Middle Pleistocene	L'Aven de Romain-La-Roche, France	Tyrberg (2008a)

(continued)

Species	Period	Location	References
F. peregrinus	Middle Pleistocene	Abîmes de La Fage, France	Mourer-Chauviré et al. (2003)
F. peregrinus	Middle/Late Pleistocene	Tropfsteinhöhle, Austria	Tyrberg (2008a)
F. peregrinus	Middle/Late Pleistocene	Binagady, Azerbaijan	Tyrberg (2008a)
F. peregrinus	Middle/Late Pleistocene	Stránská Skála, Czech Republic	Tyrberg (2008a)
F. peregrinus	Late Pleistocene	Gudenushöhle, Austria	Tyrberg (2008a)
F. peregrinus	Late Pleistocene	Tunnelhöhle, Austria	Tyrberg (2008a)
F. peregrinus	Late Pleistocene	Gorham's Cave, Gibraltar	Finlayson et al. (2016)
F. peregrinus	Late Pleistocene	Zafarraya, Spain	Finlayson et al. (2016)
F. peregrinus	Late Pleistocene	Abri Dufaure, France	Tyrberg (2008a)
F. peregrinus cyrneus†	Late Pleistocene	Grotte de la Coscia (Corse), France	Tyrberg (2008a)
F. peregrinus	Late Pleistocene	Combe Saunière 1, France	Tyrberg (2008a)
F. peregrinus	Late Pleistocene	Liko cave (Crete), Greece	Tyrberg (2008a)
F. peregrinus	Late Pleistocene	Hayonim cave (Galilee), Israel	Tyrberg (2008a)
F. peregrinus	Late Pleistocene	Cala della Calcina (Sardegna), Italy	Tyrberg (2008a)
F. peregrinus	Late Pleistocene	Grotta Romanelli (Puglia), Italy	Tyrberg (2008a)
F. peregrinus	Late Pleistocene	Brinzeni, Moldova	Tyrberg (2008a)
F. peregrinus	Late Pleistocene	Cueva de Valdegoba, Spain	Tyrberg (2008a)
F. peregrinus	Late Pleistocene	Komarowa Cave, Poland	Tyrberg (2008a)
F. peregrinus	Late Pleistocene	Oblazowa Cave, Poland	Tyrberg (2008a)
F. peregrinus	Late Pleistocene	Krucza Skala, Poland	Tyrberg (2008a)
F. peregrinus	Late Pleistocene	Tell Mureybet, Syria	Tyrberg (2008a)
F. peregrinus	Late Pleistocene	Cueva Negra de la Incarnación, Spain	Tyrberg (2008a)
F. peregrinus	Late Pleistocene	Sesselfelsgrotte, Germany	Tyrberg (2008a)
F. peregrinus	Late Pleistocene	Gouffre du Haut, France	Tyrberg (2008a)
F. peregrinus	Late Pleistocene	Emine-Bair-Khosar, Ukraine	Tyrberg (2008a)
F. peregrinus	Late Pleistocene	Grotte de Bourrouilla, France	Tyrberg (2008a)
F. peregrinus	Late Pleistocene	Loutra Almopias Cave, Greece	Boev and Tsoukala (2019)
F. peregrinus	Late Pleistocene	Hortus Herault, France	Eastham (2021)
F. peregrinus	Late Pleistocene	Soulabé, Ariège, France	Eastham (2021)
F. antiquus†	Early Pleistocene	Monte Peglia, Italy	Piperno et al. (1984)
F. antiquus†	Middle Pleistocene	Abîmes de La Fage, France	Mourer-Chauviré et al. (2003)
F. antiquus†	Middle Pleistocene	Horvolgy Cave, Hungary	Janossy (1977)
Falco sp. *(large)*	Middle Pleistocene	Stránská Skála, Czech Republic	Tyrberg (2008a)

(continued)

Species	Period	Location	References
Falco sp. *(large)*	Late Pleistocene	Prepoštská Cave, Slovakia	Sabol et al. (2013)
Falco sp. *(large)*	Late Pleistocene	Cave no. 16, Bulgaria	Tyrberg (2008a)
Falco sp. *(large)*	Late Pleistocene	Grotte de la Coscia (Corse), France	Tyrberg (2008a)

References

Altolaguirre Y, Bruch AA, Gibert L (2020) A long Early Pleistocene pollen record from Baza Basin (SE Spain): major contributions to the palaeoclimate and palaeovegetation of Southern Europe. Quat Sci Rev 231:106199. https://doi.org/10.1016/j.quascirev.2020.106199

Badino F, Pini R, Ravazzi C, Margaritora D, Arrighi S, Bortolini E et al (2020) An overview of Alpine and Mediterranean palaeogeography, terrestrial ecosystems and climate history during MIS 3 with focus on the Middle to Upper Palaeolithic transition. Quat Int 551:7–28. https://doi.org/10.1016/j.quaint.2019.09.024

Becker JJ (1987) Revision of "*Falco*" ramenta Wetmore and the Neogene evolution of the *Falconidae*. Auk 104:270–276. https://doi.org/10.1093/auk/104.2.270

Bedetti C, Pavia M (2007) Reinterpretation of the Late Pleistocene Ingarano cave deposit based on the fossil bird associations (Apulia, South-Eastern Italy). Riv Ital Paleont Strat 113:487–507. http://hdl.handle.net/2318/80201

Bedetti C, Pavia M (2013) Early Pleistocene birds from Pirro Nord (Puglia, southern Italy). Palaeontographica A 298(1–6):31–53. https://doi.org/10.1127/pala/298/2013/31

Berlusconi A, Preatoni D, Assandri A, Bisi F, Brambilla M et al (2022) Intra-guild spatial niche overlap among three small falcon species in an area of recent sympatry. Eur Zool J 89:510–526. https://doi.org/10.1080/24750263.2022.2055170

Berto C, Krajcarz MT, Moskal-del Hoyo M, Komar M, Sinet-Mathiot V, Zarecka-Szubińska K et al (2021) Environment changes during Middle to Upper Palaeolithic transition in Central Europe. A multiproxy approach for the MIS 3 sequence of Koziarnia Cave (Kraków-Częstochowa Upland, Poland). J Archaeol Sci Rep 35. https://doi.org/10.1016/j.jasrep.2020.102723

Blanco G, Sánchez-Marco A, Negro JJ (2021) Night capture of roosting cave birds by Neanderthals: an actualistic approach. Front Ecol Evol 9:733062. https://doi.org/10.3389/fevo.2021.733062

Blondel J (1985) Historical and ecological evidence on the development of Mediterranean avifaunas. In: Ilyichev VD, Gavrilov VM (eds) Acta XVIII Congressus Internationalis Ornithologici. Nauka Publishing House, Moscow, pp 373–386

Blondel J, Mourer-Chauviré C (1998) Evolution and history of the western Palearctic avifauna. Trends Ecol Evol 13(12):488–492. https://doi.org/10.1016/S0169-5347(98)01461-X

Bocheński ZM (1984) Middle Pleistocene remains of birds from Kozi Grzbiet in the Świętokrzyskie Mts. (Holy Cross Mts.—Central Poland). Acta Zool Cracov 27:177–186

Bocheński ZM (1991) Pleistocene bird fauna from Bir Tarfawi (Egyptian, Western Desert). Ostrich 62:29–34. https://doi.org/10.1080/00306525.1991.9639636

Bocheński ZM, Huhtala K, Jussila P, Pulliainen E, Tornberg R, Tunkkari PS (1998) Damage to bird bones in pellets of Gyrfalcon *Falco rusticolus*. J Archaeol Sci 25:425–433. https://doi.org/10.1006/jasc.1997.0213

Bocheński ZM, Tomek T, Tornberg R, Wertz K (2009) Distinguishing nonhuman predation on birds: pattern of damage done by the white-tailed eagle *Haliaetus albicilla*, with comments on the punctures made by the golden eagle *Aquila chrysaetos*. J Archaeol Sci 36:122–129. https://doi.org/10.1016/j.jas.2008.07.018

Boev ZN (1999) *Falco bakalovi* sp. n.—a Late Pliocene falcon (*Falconidae, Aves*) from Varshets (W Bulgaria). Geol Balc 29(1–2):131–135

Boev ZN, Tsoukala E (2019) Late Pleistocene and Earliest Holocene avifauna from the Loutra Almopias Cave (Macedonia, Greece). Hist Nat Bulg 40:1–31

Burnham K, Burnham W, Newton I (2009) Gyrfalcon *Falco rusticolus* post-glacial colonization and extreme long-term use of nest-sites in Greenland. Ibis 151:514–522. https://doi.org/10.1111/j.1474-919X.2009.00939.x

Calatayud J, Rodríguez MA, Molina-Venegas R, Leo M, Horreo JL, Hortal J (2019) Pleistocene climate change and the formation of regional species pools. Proc R Soc B 286:20190291. https://doi.org/10.1098/rspb.2019.0291

Carrera L (2018) Birds from the Late Pleistocene: environmental and climatic scenarios between the Alps and the Great Adriatic Plain (North-Eastern Italy). Alpine Medit Quat 31:49–53

Carrera L, Scarponi D, Martini F, Sarti L, Pavia M (2021) Mid-Late Pleistocene Neanderthal landscapes in Southern Italy: paleoecological contributions of the avian assemblage from Grotta del Cavallo, Apulia, southern Italy. Palaeogeogr Palaeoclimatol Palaeoecol 567:110256. https://doi.org/10.1016/j.palaeo.2021.110256

Cenizo M, Noriega JI, Reguero MA (2016) A stem falconid bird from the Lower Eocene of Antarctica and the early southern radiation of the falcons. J Ornithol 157(3):885–894. https://doi.org/10.1007/s10336-015-1316-0

Clark CD, Hughes ALC, Greenwood SL, Jordan C, Sejrup HP (2012) Pattern and timing of retreat of the last British–Irish Ice Sheet. Quat Sci Rev 44:112–146. https://doi.org/10.1016/j.quascirev.2010.07.019

Cuenca-Bescós G, Rofes J, López-García JM, Blain H-A, Rabal-Garcés R, Sauqué V, Arsuaga JL, Bermúdez de Castro JM, Carbonell E (2013) The small mammals of Sima del Elefante (Atapuerca, Spain) and the first entrance of *Homo* in Western Europe. Quat Int 295:28–35. https://doi.org/10.1016/j.quaint.2011.12.012

Dalén L, Fuglei E, Hersteinsson P, Kapel CMO, Roth JD, Samelius G, Tannerfeldt M, Angerbjörn A (2005) Population history and genetic structure of a circumpolar species: the arctic fox. Biol J Linn Soc 84:70–89. https://doi.org/10.1111/j.1095-8312.2005.00415.x

Davidson I (1989) La economía del final del Paleolítico en la España oriental. Servicio de Investigación Prehistórica 85, Diputación Provincial de Valencia, Valencia

Delpech F (1983) Les faunes du Paléolithique supérieur dans le sud-ouest de la France. Centre National de la Recherche Scientifique, Paris

Di Pasquale G, Saracino A, Bosso L, Russo D, Moroni A, Bonanomi G, Allevato E (2020) Coastal pine-oak glacial qrefugia in the Mediterranean basin: a biogeographic approach based on charcoal analysis and spatial modelling. Forests 11:673. https://doi.org/10.3390/f11060673

Dogandžić M, Đuričić L (2017) Lithic production strategies in the Middle Paleolithic of the southern Balkans. Quat Int 450:68–102. https://doi.org/10.1016/j.quaint.2017.03.011

Doyle JM, Bell DA, Bloom PH, Emmons G, Fesnock A, Katzner TE et al (2018) New insights into the phylogenetics and population structure of the prairie falcon (*Falco mexicanus*). BMC Genomics 19:233. https://doi.org/10.1186/s12864-018-4615-z

Eastham A (2021) Man and bird in the palaeolithic of Western Europe. Archaeopress Publishing, Oxford

Elorza M (1990) Bird remains in prehistoric settlements from Basque Country. Carried out investigations. Munibe (Antropologia -Arkeologia) 42:263–267

Figueiredo SD, Cunha PP, Carvalho IS (2022) Marine mammals fossil remains and synthesis of the sedimentary and paleontological record of the Furninha Cave Pleistocene (Peniche, Portugal). Cuaternario y Geomorf 36:37–52. https://doi.org/10.17735/cyg.v36i1-2.89297

Finlayson C (2011) Avian survivors. The history and biogeography of palearctic birds. Poyser, London

Finlayson S (2019) Birds in the Neanderthal ecological niche—using birds from Pleistocene cave archives to characterise the relationship between Neanderthals and birds. PhD thesis, Anglia Ruskin University

Finlayson C, Carrión J, Brown K, Finlayson G, Sánchez-Marco A, Fa D, Rodríguez-Vidal J, Fernández S, Fierro E, Bernal-Gómez M, Giles-Pacheco F (2011) The Homo habitat niche: using the avian fossil record to depict ecological characteristics of Palaeolithic Eurasian hominins. Quat Sci Rev 30:1525–1532. https://doi.org/10.1016/j.quascirev.2011.01.010

Finlayson C, Brown K, Blasco R, Rosell J, Negro JJ et al (2012a) Birds of a feather: neanderthal exploitation of raptors and corvids. PLoS ONE 7(9):e45927. https://doi.org/10.1371/journal.pone.0045927

Finlayson C, Monclova A, Carrión JS, Fa DA, Finlayson G, Rodríguez-Vidal J, Giles-Pacheco F (2012b) Ecological transitions—but for whom? A perspective from the Pleistocene. Palaeogeogr Palaeoclimatol Palaeoecol 329–330:1–9. https://doi.org/10.1016/j.palaeo.2011.04.002

Finlayson C, Finlayson S, Giles Guzman F, Sanchez Marco A, Finlayson G, Jennings R, Giles Pacheco F, Rodriguez Vidal J (2016) Using birds as indicators of Neanderthal environmental quality: Gibraltar and Zafarraya compared. Quat Int 421:32e45. https://doi.org/10.1016/j.quaint.2015.11.031

Fuchs J, Johnson JA, Mindell DP (2015) Rapid diversification of falcons (*Aves*: *Falconidae*) due to expansion of open habitats in the Late Miocene. Mol Phylogenet Evol 82:166–182. https://doi.org/10.1016/j.ympev.2014.08.010

Gál E (2008) Faunal and taphonomic analyses of a Late Pleistocene bird-bone assemblage from a cave deposit in north-west Hungary. Geobios 41:79–90. https://doi.org/10.1016/j.geobios.2006.06.005

Gál E, Dimitrijevic V, Kessler E (2003) Bird remains from the Late Pleistocene deposits of Mališhina Stijena (Montenegro). Theor Appl Karstol 16:67–75

Graham AGC, Stoker MS, Lonergram L, Bradwell T, Stewart MA (2011) The Pleistocene glaciations of the North Sea Basin. In: Ehlers J, Gibbard PL, Hughes PD (eds) Quaternary glaciations extent and chronology: a closer look, Developments in quaternary science, vol 15. Elsevier, Amsterdam, pp 261–278

Griffiths CS (1999) Phylogeny of the *Falconidae* inferred from molecular and morphological data. Auk 116:116–130. https://doi.org/10.2307/4089459

Groombridge JJ, Jones CG, Bayes MK, van Zyl AJ, Carrillo J, Nichols RA, Bruford MW (2002) A molecular phylogeny of African kestrels with reference to divergence across the Indian Ocean. Mol Phylogenet Evol 25(2):267–277. https://doi.org/10.1016/S1055-7903(02)00254-3

Gu Z, Pan S, Lin Z, Hu L, Dai X, Chang J, Xue Y, Su H, Long J, Sun M, Ganusevich S, Sokolov V, Sokolov A, Pokrovsky I, Ji F, Bruford MW, Dixon A, Zhan X (2021) Climate-driven flyway changes and memory-based long-distance migration. Nature 591:259–264. https://doi.org/10.1038/s41586-021-03265-0

Guerra C, McMinn M, Alcover JA (2013) The Upper Pleistocene-Holocene raptorial bird guild from Eivissa Island (Pityusic Archipelago, Western Mediterranean Sea). Geobios 46:491–502. https://doi.org/10.1016/j.geobios.2013.10.003

Hardy BL (2010) Climatic variability and plant food distribution in Pleistocene Europe: implications for Neanderthal diet and subsistence. Quat Sci Rev 29:662–679. https://doi.org/10.1016/j.quascirev.2009.11.016

Harrison CJO (1980) Pleistocene bird remains from Tornewton Cave and the Brixham Windmill Hill Cave in south Devon. Bull Br Mus Nat Hist (Geol) 33:91–100

Harvati K, Darlas A, Bailey SE, Rein TR, El Zaatari S, Fiorenza L, Kullmer O, Psathi E (2013) New Neanderthal remains from Mani peninsula, southern Greece: the Kalamakia middle Paleolithic cave site. J Hum Evol 64:486–499. https://doi.org/10.1016/j.jhevol.2013.02.002

Holder K, Montgomerie R, Friesen VL (1999) A test of the glacial refugium hypothesis using patterns of mitochondrial and nuclear DNA sequence variation in Rock Ptarmigan (*Lagopus mutus*). Evolution 53:1936–1950. https://doi.org/10.1111/j.1558-5646.1999.tb04574.x

Holm SR, Svenning J-C (2014) 180,000 years of climate change in Europe: avifaunal responses and vegetation implications. PLoS ONE 9(4):e94021. https://doi.org/10.1371/journal.pone.0094021

Holman JA (1998) Pleistocene amphibians and reptiles in Britain and Europe. Oxford University Press, Oxford

Houmark-Nielsen M (2011) Pleistocene glaciations in Denmark: a closer look at chronology, ice dynamics and landforms. In: Ehlers J, Gibbard PL, Hughes PD (eds) Quaternary glaciations—extent and chronology: a closer look, Developments in quaternary science, vol 15. Elsevier, Amsterdam, pp 47–58

Hughes PD, Gibbard PL, Ehlers J (2013) Timing of glaciation during the last glacial cycle: evaluating the concept of a global 'Last Glacial Maximum'(LGM). Earth-Sci Rev 125:171–198. https://doi.org/10.1016/j.earscirev.2013.07.003

Huguet R, Vallverdú J, Rodríguez-Álvarez XP, Terradillos-Bernal M, Bargalló A, Lombera-Hermida A, Menéndez L, Modesto-Mata M, Van der Made J, Soto M, Blain HA, García N, Cuenca-Bescós G, Gómez-Merino G, Pérez-Martínez R, Expósito I, Allué E, Rofes J, Burjachs F, Canals A, Bennàsar M, Nuñez-Lahuerta C, Bermúdez de Castro JM, Carbonell E (2017) Level TE9c of Sima del Elefante (Sierra de Atapuerca, Spain): a comprehensive approach. Quat Int 433A:278–295. https://doi.org/10.1016/j.quaint.2015.11.030

Huntley B, Green RE (2011) Bioclimatic models of the distributions of gyrfalcons and ptarmigan. In: Watson RT, Cade TJ, Fuller M, Hunt G, Potapov E (eds) Gyrfalcons and ptarmigan in a changing world, vol II. The Peregrine Fund, Boise, pp 329–338. https://doi.org/10.4080/gpcw.2011.0312

Jánossy D (1972) Die mittelpleistozäne Vogelfauna der Stránská Skála. Anthropos 21:35–64

Jánossy D (1977) Plio-Pleistocene birds remains from the Carpathian basin III. Strigiformes, falconiformes, caprimulgiformes, apodiformes. Aquila 84:9–36

Jenkinson RDS, Gilbertson DD, Sutherland S (1984) Changes in European geese and duck migration patterns during the Quaternary. In: Jenkinson RDS, Gilbertson DD (eds) In the shadow of extinction: a quaternary archaeology and palaeoecology of the lake, fissures and smaller caves at Cresswell Crags SSSI. Nottinghamshire and Derbyshire County Councils

Johnsen A, Rindal E, Ericson PGP, Zuccon D, Kerr KCR, Stoeckle MY, Lifjeld JT (2010) DNA barcoding of Scandinavian birds reveals divergent lineages in trans-Atlantic species. J Ornithol 151:565–578. https://doi.org/10.1007/s10336-009-0490-3

Johnson JA, Burnham KK, Burnham WA, Mindell DP (2007) Genetic structure among continental and island populations of gyrfalcons. Mol Ecol 16(15):3145–3160. https://doi.org/10.1111/j.1365-294X.2007.03373.x

Joseph S, O'Connor RE, Al Mutery AF, Watson M, Larkin DM, Griffin DK (2018) Chromosome level genome assembly and comparative genomics between three falcon species reveals an unusual pattern of genome organisation. Diversity 10:113. https://doi.org/10.3390/d10040113

Kangas V-M, Carrillo J, Debray P, Kvist L (2018) Bottlenecks, remoteness and admixture shape genetic variation in island populations of Atlantic and Mediterranean common kestrels *Falco tinnunculus*. J Avian Biol 49:e01768. https://doi.org/10.1111/jav.01768

Karyakin IV (2011) Subspecies population structure of the Saker falcon range. Raptor Conserv 21:116–172

Kessler J (2009) New results with regard to the Neogene and Quaternary Avifauna of the Carpathian basin. Part I. Földtani Közlöny 139:67–82

Kessler J (2018) Evolution and presence of diurnal predatory birds in the Carpathian Basin. Ornis Hung 26:102–123. https://doi.org/10.1515/orhu-2018-0008

Laroulandie V (2002) Damage to pigeon long bones in pellets of the eagle owl *Bubo bubo* and food remains of peregrine falcon *Falco peregrinus*: zooarchaeological implications. Acta Zool Cracov 45:331–339

Laroulandie V (2005) Anthropogenic versus non-anthropogenic bird bone assemblages: new criteria for their distinction. In: O'Connor T (ed) Biosphere to lithosphere. New studies in vertebrate taphonomy. Oxbow Books, Oxford, pp 25–30

Laroulandie V (2009) De la plume à l'œuf: exploitation des ressources aviaires au Magdalénien dans le Sud de la France. In: Fontana L, Chauvière FX, Bridault A (eds) In search of total

animal exploitation, Cases studies in upper palaeolithic and mesolithic. J. & E. Hedges, Oxford, pp 71–89
Leonardi G (2015) The Lanner falcon. GLE, Catania
Leonardi G (2020) Behavioural ecology of Western Palearctic falcons. Springer, Cham. https://doi.org/10.1007/978-3-030-60541-4
Li Z, Zhou Z, Deng T, Li Q, Clarke JA (2014) A falconid from the Late Miocene of northwestern China yields further evidence of transition in Late Neogene steppe communities. Auk 131(3):335–350. https://doi.org/10.1642/AUK-13245.1
Lloveras L, Garcia L, Maroto J, Soler J, Soler N (2018) The bird assemblage from the Middle Palaeolithic level I of Arbreda Cave: a taphonomic story. J Archaeol Sci Rep 21:758–770. https://doi.org/10.1016/j.jasrep.2018.08.040
Louchart A (2002) Les oiseaux du Pléistocene de Corse et de quelques localités sardes. Ecologie, evolution, biogeography et ectinction. Docum Lab Geol Lyon 155:1–287
Louchart A (2011) Aves. In: Harrison TH (ed) Paleontology and geology of Laetoli: human evolution in context. Volume 2: fossil hominins and the associated fauna. Vertebrate Paleobiology and Paleoanthropology Series. Springer, New York, pp 505–533. https://doi.org/10.1007/978-90-481-9962-4_18
Martínez Valle R, Guillem Calatayud PM, Villaverde Bonilla V (2016) Bird consumption in the final stage of Cova Negra (Xátiva, Valencia). Quat Int 421:85–102. https://doi.org/10.1016/j.quaint.2016.01.068
McCarty JP (2001) Ecological consequences of recent climate change. Conserv Biol 15:320–331. https://doi.org/10.1046/j.1523-1739.2001.015002320.x
Mihailović D, Milošević S, Blackwell BAB, Mercier N, Mentzer SM, Miller CE, Morley MW et al (2022) Neanderthal settlement of the Central Balkans during MIS 5: evidence from Pešturina cave, Serbia. Quat Int 610:1–19. https://doi.org/10.1016/j.quaint.2021.09.003
Milliken S (1991) Aspects of lithic assemblage variability in the Late Palaeolithic of South-East Italy. PhD thesis, Oxford University, Oxford
Mlíkovský J (2002) Cenozoic birds of the world. Part 1: Europe. Ninox Press, Praha
Mlíkovský J (2009) Middle Pleistocene birds of Hundsheim, Austria. Jour Nat Mus Prague (Nat Hist S) 177:69–82
Moreno-Garcia M, Pimenta CM (2002) The Paleofaunal context. In: Zilhão J, Trinkaus E (eds) Portrait of the artist as a child: the Gravettian human skeleton from the Abrigo do Lagar Velho and its archaeological context, Trabalhos de Arqueologia, vol 22. Instituto Portugûes de Arqueologia, Lisboa, pp 112–131
Morin E, Laroulandie V (2016) Presumed symbolic use of diurnal raptors by Neanderthals. PLoS ONE 7:e32856. https://doi.org/10.1371/journal.pone.0032856
Mourer-Chauviré C (1975) Les oiseaux du Pléistocène moyen et supérieur de France. Doc Lab Geol Fac Sc Lyon 64:1–624
Mourer-Chauviré C (1993) The Pleistocene avifaunas of Europe. Archaeofauna 2:53–66
Mourer-Chauviré C, Philippe M, Quinif Y, Chaline J, Dabard E, Guérin C, Hugueney M (2003) Position of the palaeontological site Aven I des Abîmes de La Fage, at Noailles (Corrèze, France), in the European Pleistocene chronology. Boreas 32:521–531. https://doi.org/10.1111/j.1502-3885.2003.tb01232.x
Murray J, Domínguez-Alonso P, Fernádez-Jalvo Y, King T, Lynch EP, Andrews P, Yepiskoposyan L (2010) Pleistocene to Holocene stratigraphy of Azokh 1 Cave, Lesser Caucasus. Irish J Earth Sci 28:75–91. https://doi.org/110.1353/ijes.2010.a810034
Negro JJ, Prenda J, Ferrero JJ, Rodríguez A, Reig-Ferrer A (2020) A timeline for the urbanization of wild birds: the case of the lesser kestrel. Quat Sci Rev 249:106638. https://doi.org/10.1016/j.quascirev.2020.106638
Newton I, Dale L (2001) A comparative analysis of the avifaunas of different zoogeographical regions. J Zool 254(2):207–218. https://doi.org/10.1017/S0952836901000723

Nittinger F, Haring E, Pinsker W, Wink M, Gamauf A (2005) Out of Africa? Phylogenetic relationships between *Falco biarmicus* and the other Hierofalcons (*Aves*: *Falconidae*). J Zool Syst Evol Res 43(4):321–331. https://doi.org/10.1111/j.1439-0469.2005.00326.x
Nittinger F, Gamauf A, Pinsker W, Wink M, Haring E (2007) Phylogeography and population structure of the saker falcon (*Falco cherrug*) and the influence of hybridization: mitochondrial and microsatellite data. Mol Ecol 16(7):1497–1517. https://doi.org/10.1111/j.1365-294X.2007.03245.x
Núñez-Lahuerta C, GalÁn J, Sauqué V, Cuenca-Bescós G (2018) The role of birds in Late Pleistocene Eurosiberian Mediterranean boundary reconstructions in Western Europe. Quat Int 481:101–112. https://doi.org/10.1016/j.Quaint.2018.01.023
Núñez-Lahuerta C, Galán J, Cuenca-Bescós G, Huguet R (2021) Birds from sima del elefante, atapuerca, spain: palaeoecological implications in the oldest human bearing levels of the iberian Peninsula. Riv It Paleontol Strat 127:453–484. https://doi.org/10.13130/2039-4942/16027
Nyström J, Dalén L, Hellström P, Ekenstedt J, Angleby H, Angerbjörn A (2006) Effect of local prey availability on gyrfalcon diet: DNA analysis on ptarmigan remains at nest sites. J Zool 269:57–64. https://doi.org/10.1111/j.1469-7998.2006.00050.x
Olson SL (1985) The fossil record of birds. In: King JR, Parker DC (eds) Avian biology, vol VIII. Academic, New York, pp 79–252
Oros Sršen A, Brajković D, Radović S, Mauch Lenardić J, Miracle PT (2014) The Avifauna of Southern Istria (Croatia) during the late Pleistocene: implications for the palaeoecology and biodiversity of the Northern Adriatic region. Int J Osteoarchaeol 24(3):289–299. https://doi.org/10.1002/oa.2364
Pan S, Zhang T, Rong Z, Hu L, Gu Z, Wu Q et al (2017) Population transcriptomes reveal synergistic responses of DNA polymorphism and RNA expression to extreme environments on the Qinghai Tibetan Plateau in a predatory bird. Mol Ecol 26(11):2993–3010. https://doi.org/10.1111/mec.14090
Pande S, Yosef R, Morelli F, Pawar R, Mone R (2018) Diet and habitat affinities in six raptor species in India. Avian Res 9(36). https://doi.org/10.1186/s40657-018-0129-2
Peresani M, Fiore I, Gala M, Romandini M, Tagliacozzo A (2011) Late Neandertals and the intentional removal of feathers as evidenced from bird bone taphonomy at Fumane Cave 44 ky B.P., Italy. Proc Natl Acad Sci 108:3888–3893. https://doi.org/10.1073/pnas.1016212108
Pfeffer R (2012) The role of hybridisation in origin of forms in the *Hierofalco* complex. Raptor Conserv 24:148–164
Piperno M, Segre AG, Segre Naldini E (1984) Catalogue: I primi abitanti d'Europa. De Luca Ed, Monte Peglia, Umbria, pp 115–119
Porter D, White CM (1973) The Peregrine Falcon in Utah, emphasizing ecology and competition with the Prairie Falcon. BrighamYoung Univ Sd Bull 18:1–74
Potapov E, Sale R (2005) The Gyrfalcon. T & AD Poyser, London
Ray N, Adams JM (2001) A GIS-based vegetation map of the world at the last glacial maximum (25,000–15,000 BP). Internet Archaeol 11. http://intarch.ac.uk/journal/issue11/rayadams_toc.html
Richards MP, Trinkaus E (2009) Out of Africa: modern human origins special feature: isotopic evidence for the diets of European Neanderthals and early modern humans. Proc Natl Acad Sci USA 106:16034–16039. https://doi.org/10.1073/pnas.0903821106
Romandini M, Fiore I, Gala M, Cestari M, Guida G, Tagliacozzo A, Peresani M (2016) Neanderthal scraping and manual handling of raptors wing bones: evidence from Fumane Cave, experimental activities and comparison. Quat Int 421:154–172. https://doi.org/10.1016/j.quaint.2015.12.078
Rufà A, Laroulandie V (2021) Unravelling the taphonomic stories of bird bones from the Middle Pleistocene Layer VIII of Grotte Vaufrey, France. Quat J 4:30. https://doi.org/10.3390/quat4040030
Sabol M, Čeklovský T, Beňuš R, Kováčová M, Joniak P, Zervanová J, Putiška R (2013) Fossil assemblages from Neanderthal sites in Slovakia—preliminary results. In: Filippi M, Bosák P (eds) Proceedings of the 16th International Congress of Speleology, Vol 1. Czech Speleological Society, Praha, pp 192–195

Sánchez-Marco A (1999) Implications of the avian fauna for paleoecology in the Early Pleistocene of the Iberian Peninsula. J Hum Evol 37:375–388. https://doi.org/10.1006/jhev.1999.0345
Sánchez-Marco A (2004) Avian zoogeographical patterns during the Quaternary in the Mediterranean Region and paleoclimatic interpretation. Ardeola 51:91–132
Šedivá M, Obuch J, Sabol M (2020) Avifauna from the Neanderthal Site of Prepoštská jaskyňa Cave (Bojnice I) near Prievidza—biostratigraphical and palaeoecological implications. Acta Geol Slovaca 12:57–68
Simmons T, Nadel D (1998) The avifauna of the Early Epipalaeolithic site of Ohalo 2 (19400 years BP), Israel: species diversity, habitat and seasonality. Int J Osteoarchaeol 8:79–96. https://doi.org/10.1002/(SICI)1099-1212(199803/04)8:23.0.CO;2-I
Speth JD, Spielmann KA (1983) Energy source, protein metabolism, and hunter-gatherer subsistence strategies. J Anthropol Archaeol 2:1–31. https://doi.org/10.1016/0278-4165(83)90006-5
Stewart JR (2007) The fossil and archaeological record of the Eagle Owl in Britain. Brit Birds 100:481–486
Stewart JR, Lister AM, Barnes I, Dalén L (2010) Refugia revisited: individualistic responses of species in space and time. Proc R Soc B Biol Sci 277:661–671. https://doi.org/10.1098/rspb.2009.1272
Sveinsdóttir M, Guðmundsdóttir L, Magnússon KP (2017) Complete mitochondrial genome of the gyrfalcon *Falco rusticolus* (Aves, Falconiformes, Falconidae). Mitochondrial DNA Part A 28:370–371. https://doi.org/10.3109/19401736.2015.1126827
Taberlet P, Fumagalli L, Wust-Saucy A-G, Cosson J-F (1998) Comparative phylogeography and postglacial colonization routes in Europe. Mol Ecol 7:453–464. https://doi.org/10.1046/j.1365-294x.1998.00289.x
Talbot S, Palmer AG, Sage GK, Sonsthagen SA, Swem T, Brimm DJ et al (2011) Lack of genetic polymorphism among peregrine falcons of Fiji. J Av Biol 42:1–14. https://doi.org/10.1111/j.1600-048X.2011.05280.x
Tchernov E (1968) Peregrine Falcon and Purple Gallinule of late Pleistocene age in the Sudanese Aswan Reservoir area. Auk 85:133
Tomek T, Bocheński ZM, Socha P, Stefaniak K (2012) Continuous 300,000-year fossil record: changes in the ornithofauna of Biśnik Cave, Poland. Palaeontol Electron 15:2A–20p. https://doi.org/10.26879/277
Tyrberg T (1991) Arctic, montane and steppe birds as glacial relicts in the West Palearctic. Ornithol Verhandlungen 25:29–49
Tyrberg T (2008a) Pleistocene birds of the Palearctic. http://web.telia.com/~u11502098/pleistocene.pdf
Tyrberg T (2008b) The Late Pleistocene Continental Avian extinction—an evaluation of the fossil evidence. Oryctos 7:249–269
Tyrberg T (2010) Avifaunal responses to warm climate: the message from Last Interglacial Faunas. Rec Aust Mus 62:193–205. https://doi.org/10.3853/j.0067-1975.62.2010.1543
Umanskaya AS (1981) The Miocene birds of the Western Black Sea Coasts of the Ukrainian SSR. Vestn Zool 17:17–21
Urban B, Sierralta M, Frechen M (2011) New evidence for vegetation development and timing of Upper Middle Pleistocene interglacials in Northern Germany and tentative correlations. Quat Int 241:125–142. https://doi.org/10.1016/j.quaint.2011.02.034
Waltari E, Cook JA (2005) Hares on ice: phylogeography and historical demographics of *Lepus arcticus*, *L. othus*, and *L. timidus* (Mammalia: Lagomorpha). Mol Ecol 14:2005–3016. https://doi.org/10.1111/j.1365-294X.2005.02625.x
White CM, Sonsthagen SA, Sage GK, Anderson C, Talbot SL (2013) Genetic relationships among some subspecies of the Peregrine falcon (*Falco peregrinus* L.), inferred from mitochondrial DNA control-region sequences. Auk 130(1):78–87. https://doi.org/10.1525/auk.2012.11173
Wink M, Sauer-Gürth H (2004) Phylogenetic relationships in diurnal raptors based on nucleotide sequences of mitochondrial and nuclear marker genes. In: Chancelor RD, Meyburg B-U (eds) Raptors worldwide. WWGBP, Berlin, pp 483–498

Winn Jones R (2006) Applied palaeontology. Cambridge University Press, Cambridge

Yang C, Yang M, Wang Q, Lu Y, Li X (2018) The complete mitogenome of *Falco amurensis* (Falconiformes, Falconidae), and a comparative analysis of genus *Falco*. Zool Sci 35(4):367–372. https://doi.org/10.2108/zs170182

Yll R, Carrión JS, Marra AC, Bonfiglio L (2006) Vegetation on the basis of pollen in Late Pleistocene hyena coprolites form San Teodoro Cave (Sicily, Italy). Palaeogeogr Palaeoclimatol Palaeoecol 237:32–39. https://doi.org/10.1016/j.palaeo.2005.11.027

Zelenkov NV (2013) Zoological problems of Quaternary paleornithology. Zool Zh 92:1077–1087

Zelenkov NV (2017) Revision of non-passeriform birds from Polgárdi (Hungary, Late Miocene): 3. Neoaves. Paleontol J 51:203–213. https://doi.org/10.1134/S0031030117020162

Zhan XJ, Pan S, Wang J, Dixon A, He J et al (2013) Peregrine and saker falcon genome sequences provide insights into evolution of a predatory lifestyle. Nat Genet 45:563–566. https://doi.org/10.1038/ng.2588

Zhan XJ, Dixon A, Batbayar N, Bragin E et al (2015) Exonic versus intronic SNPs: contrasting roles in revealing the population genetic differentiation of a widespread bird species. Heredity 114(1):1–9. https://doi.org/10.1038/hdy.2014.59

Chapter 2
Holocene

2.1 The Pleistocene–Holocene Transition (PHT)

The Pleistocene–Holocene transition (PHT), which occurred between 14,000 and 10,500 BP, is characterized by rapid warming from the late Pleistocene glaciation to the early Holocene interglacial temperature levels (Table 2.1; Mol 2008; Stivrins et al. 2016; Lloveras et al. 2020). This transition resulted in global deglaciation, sea-level rise, and climate instability in the North Atlantic, Europe, and Greenland (Naughton et al. 2016). In Denmark, Germany, Poland, and Russia, rapid continental deglaciation began around 17,000, and nearly all of Britain was affected by this by 13,000 (Mol 2008). By this time, Britain was completely deglaciated, and the glacial extent had reached the middle of the Baltic (Fig. 2.1; Tomek and Gumiński 2003; Mol 2008). During the Younger Dryas, glacial advances occurred in Scandinavia and Britain, and with continuous permafrost in Poland, Germany, the Netherlands, Southern England, and Belgium, while Northern France experienced discontinuous permafrost. In the Mediterranean, a cold and dry climate prevailed, with winter precipitation in northern and central Western Iberia (Naughton et al. 2016). Between 20,000 and 18,000 BP, rainfall in North Africa became increasingly irregular, but hyper-aridity was apparent between 12,000 and 10,000 BP (Rognon 1987). In the Levant, a pronounced cold and dry episode associated with the Last Glacial Maximum (LGM) was followed by a period of warmer and wetter climate during the Bølling/Allerød phase and then a swift return to cooler and drier conditions during the Younger Dryas. This was followed by an abrupt transition to the warmer and wetter early Holocene, marking the beginning of the Epipaleolithic in this region due to the climatic improvement associated with the change from the Pleistocene to the Holocene (Lloveras et al. 2020).

During the Bølling–Allerød interstadial complex, rising temperature caused the ice sheets to continue melting (Mol 2008). This melting of the terrestrial ice sheet in Scandinavia led to the partial replacement of Arctic species with more temperate

G. Leonardi, *The History of Western Palearctic Falcons*,
https://doi.org/10.1007/978-3-031-86528-2_2

Table 2.1 Chrono-cultural, climato-stratigraphic stages, along with paleoenvironmental sequences, during the Pleistocene–Holocene transition (PHT)

Epoch	Stage	Climato-stratigraphic stage	Dates (BP)	Material culture	Phases
Holocene	Early	Boreal (BO)	10,085–8835	Epipaleolithic	Warm (mixed forests)
		Preboreal (PB)	11,490–10,085		
Pleistocene	Late Glacial	Younger Dryas	12,900–11,650	Magdalenian	Stadial (glacial tundra)
		Allerød	13,900–12,900		Interstadial (mixed forests)
		Older Dryas[a]	14,100–13,900	Epigravettian Solutrean	Stadial (Arctic tundra)
		Bølling	14,700–14,100		Interstadial (hard- and softwoods)
		Oldest Dryas	18,500–14,700		Stadial (tundra and grassland)

[a]Strongest effects, primarily in northern Europe

fauna (Fig. 2.1; Tomek and Gumiński 2003; Lambeck et al. 2010; Breivik 2014). However, cold-tolerant species persisted in the far north and the Alps (Breivik 2014). For example, the earliest known gyrfalcon, dating back to 13,000 BP (Magdalenian), was discovered in the Rhône River between Geneva and Lyon (Debrosse and Mourer-Chauviré 1973).

Climate change has significantly impacted vegetation distribution, altering the location, diversity, and accessibility of resources for animals (Rosen and Rivera-Collazo 2012). Consequently, these changes in climate and vegetation likely influenced animal behavior (Rosen and Rivera-Collazo 2012). During the Bølling–Allerød interstadial and the Younger Dryas stadial, animal communities experienced low turnover and remained relatively stable due to minimal summer temperature changes (Naughton et al. 2016; Stivrins et al. 2016; Table 2.1). The most significant rates of terrestrial and aquatic community turnover occurred during the Younger Dryas–Holocene transition around 11,700 years ago, reflecting an abrupt warming event (Stivrins et al. 2016; Lloveras et al. 2020). This period saw changes in species assemblages, including the disappearance of Arctic–alpine animals such as mammoths (Stivrins et al. 2016). Ultimately, this warming period persisted for several centuries into the early Holocene (Lloveras et al. 2020).

2.2 Early Holocene Climate in the Western Palearctic

During the Early Holocene, Eurasia experienced rapid permafrost degradation and woodland expansion. Additionally, sea levels rose by almost 60 m between 12,000 and 7000 years ago (EHSLR = Early Holocene Sea Level Rise). During the Boreal

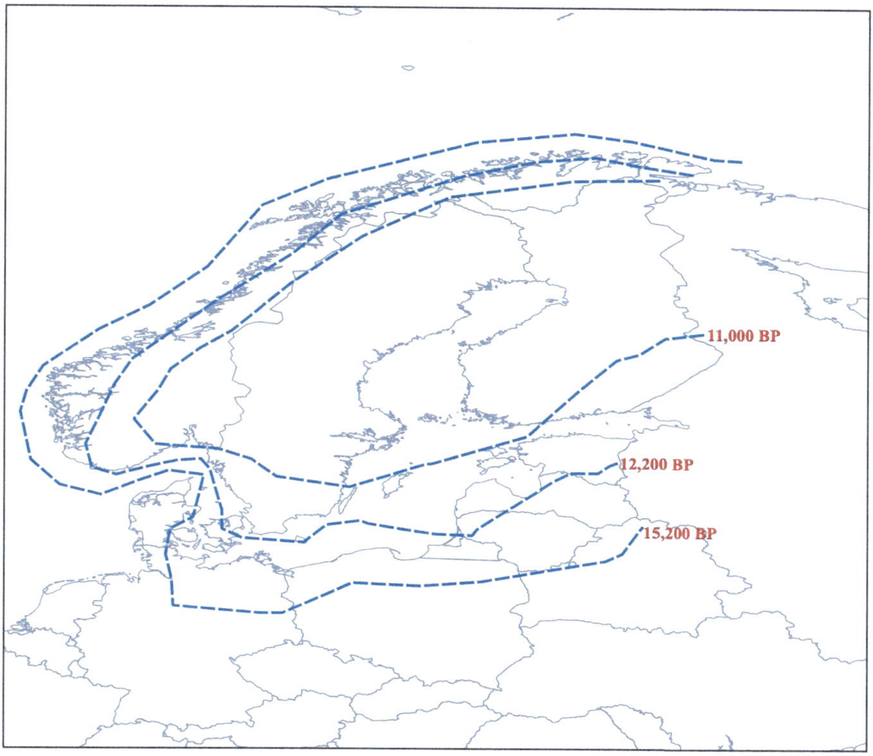

Fig. 2.1 Ice cap retreat in Scandinavia and the Baltic region during the Pleistocene–Holocene transition. (Redrawn from Tomek and Gumiński 2003)

period (9000–8000 years ago), Sweden's landscape was dominated by forests, leading to a decline in open landscape species such as the common kestrel, which likely immigrated during this time (Breivik 2014). Although a cold episode (9300–9200 BCE) led to glacial readvance, an environmental shift around 8800 BCE created more oceanic conditions. Overall, the Holocene Climate Optimum (HCO), a warm period that occurred between 9000 and 5000 years ago, brought warming to northwest of Europe, while southern Europe experienced cooling (Davis et al. 2003). The climate in North Africa became wetter, and the Sahara was dotted with numerous lakes (Rognon 1987).

The Holocene commenced in the Levant, with temperatures and rainfall levels comparable to those of today and gradually became warmer and more humid, achieving a long-term optimum around 10,500 BP (Rosen and Rivera-Collazo 2012).

2.3 Fossil Assemblages

The evolution and development of Holocene vertebrate populations in boreal regions are unfortunately poorly documented, largely due to the generally poor preservation of bones at open-air sites (Driver and Hobson 1992). Despite this, archaeological assemblages from various locations during the PHT strongly suggest a significant shift in boreal and temperate avifauna (Ericson and Tyrberg 2004; von den Driesch and Pöllath 2010). Climate changes, which impacted vegetation (particularly through reforestation), facilitated the spread of Mediterranean taxa into the northern and northwestern regions of Europe (Hewitt 1999; Davis et al. 2003; Kalis et al. 2003; Renssen et al. 2009; Stewart et al. 2010).

Overall, bird species abundance in Holocene fossil and zooarchaeological assemblages closely mirrors that of modern-day species (Turvey and Blackburn 2011; Kessler 2014; Crees et al. 2019). For instance, the ratios of all four native falcon species in the United Kingdom were similar during the Mesolithic period (Yalden and Albarella 2009). However, local variations emerged due to habitat changes, particularly following the expansion of open landscapes (Yalden and Albarella 2009). Consequently, populations of the common kestrel and merlin increased, while woodland species, such as the Eurasian hobby, declined (see Fig. 2.2).

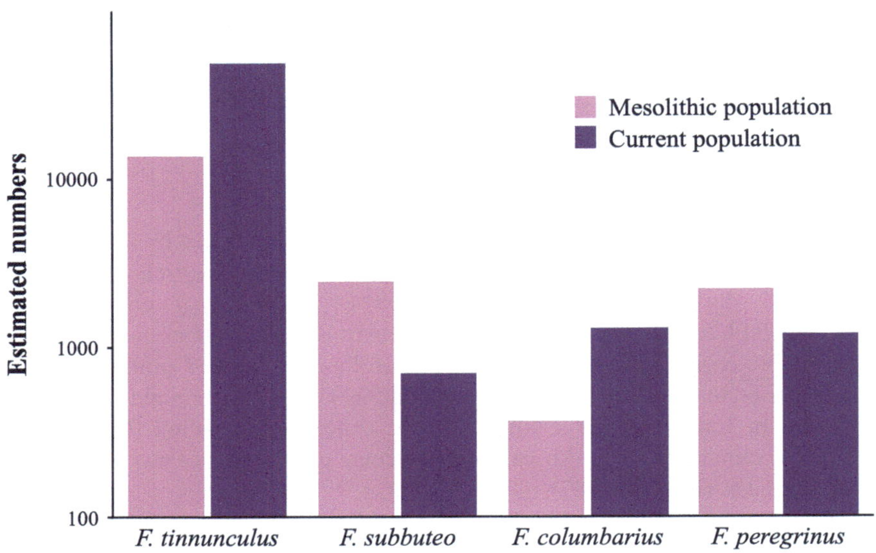

Fig. 2.2 Estimated numbers (log-transformed) of falcons in Mesolithic Britain compared to current populations. (Data from Yalden and Albarella 2009)

2.4 Human Exploitation

Falcons were likely not hunted by humans due to their lack of edibility (Warren 2015; see also Sect. 1.5). Nonetheless, zooarchaeological assemblages provide the most extensive faunal records for this period, as human populations expanded significantly during the Holocene following the Neolithic agricultural revolution (Turvey and Blackburn 2011; Appendix). Falcons may have nested in rock shelters or been preyed upon by carnivores such as wolves, foxes, weasels, martens, and lynxes, as well as avian predators such as eagle owls and eagles. Although cultures from the Late Glacial (Solutrean, Epigravettian, and Magdalenian; Sect. 2.4.1) and the Early Holocene (Epipaleolithic; Sect. 2.4.2) frequently depicted falcons in their art, these representations were likely symbolic rather than indicative of hunting practices (see also Sect. 1.5.1 for the Pleistocene).

2.4.1 Solutrean, Epigravettian, and Magdalenian Industries

Falcon fossils associated with the Solutrean (22,000–17,000 BP), Epigravettian (20,000–10,000 BP), and Magdalenian (17,000–12,000 BP) cultures, representing the last cultures of the Late Paleolithic, have been discovered at multiple locations (see Fig. 2.3 and Appendix). The predicted distribution of these prehistoric human groups aligns closely with the southern extent of periglacial environments and permafrost conditions during the Last Glacial Maximum (LGM; Banks et al. 2008).

The Solutrean culture of the Upper Paleolithic, which coincided with the LGM, was predominantly located in what is now France, Spain, and Portugal (Straus 2015). During this period, Solutrean populations exploited colder and more humid environments, corresponding to regions with permanent permafrost (Banks et al. 2008). However, fossilized remains of common and lesser kestrels have been found at several archaeological sites in southern Spain that served as glacial refugia (Fig. 2.3; Appendix). At the Solutrean level of Combe Saunière (Dordogne, France), 12,000 recognizable faunal remains were discovered, including those of common kestrel and peregrine falcon (Castel et al. 2006). These species were more adaptable to the discontinuous permafrost conditions that characterized the site (van Vliet-Lanoë et al. 2019). Bone examination indicated that the remains were handled as carrion by cave-dwelling animals (Laroulandie 2000). The presence of falcons at this site might be linked to symbolic practices involving the collection of specific birds (Castel et al. 2006).

The Epigravettian culture emerged after the LGM and spread across Italy, the Balkans, the Caucasus, Ukraine, and Western Russia. Populations from this period seem to have adapted well to regions with discontinuous or absent permafrost and seasonal freezing (Banks et al. 2008). Warm falcon species, such as the lesser kestrel and Eleonora's falcon, as well as woodland species like the Eurasian hobby and

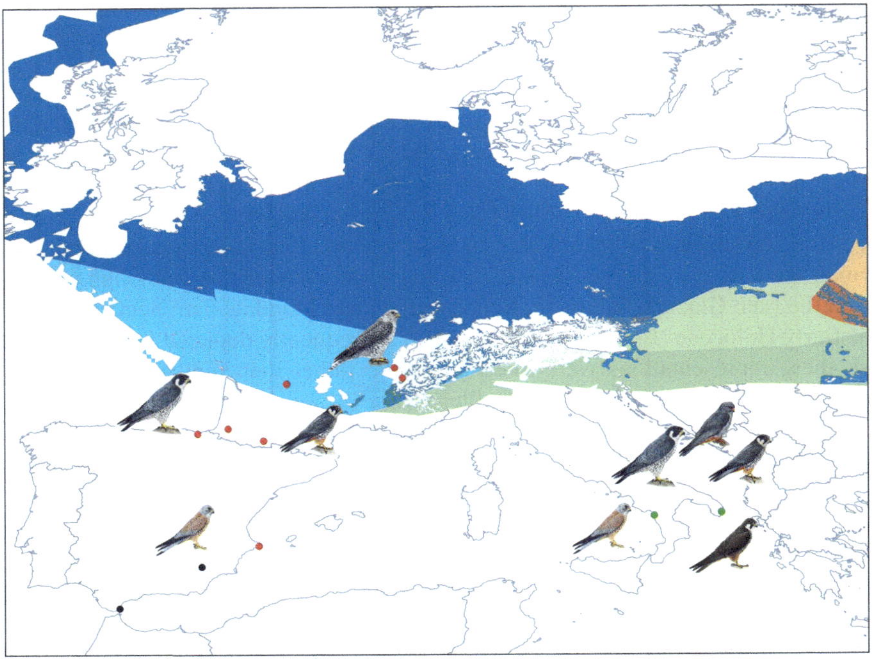

Fig. 2.3 Distribution of falcon fossils from the Solutrean (black dots), Epigravettian (green dots), and Magdalenian (red dots) industries across various archaeological sites (see Appendix; Bortolini et al. 2021). The biomes of the Last Glacial Maximum permafrost region depicted according to Lindgren (2018). Tundra—discontinuous permafrost = light blue; tundra—continuous permafrost = blue; steppe—discontinuous permafrost = green; steppe—continuous permafrost = light green

red-footed falcon, have been identified at archaeological sites in southern Italy and Montenegro (Fig. 2.3; Tyrberg 2008; Gala et al. 2016; Appendix).

The Magdalenian represents the final tradition of the European Upper Paleolithic primarily found in France, Spain, Belgium, and Germany (Fig. 2.3; Enloe 2001). During this period, the climate was predominantly cool to cold, with steppe tundra and open grasslands being the most prevalent biomes, although there were also some sparsely forested zones (Enloe 2001). As a result, warm-adapted falcon species, such as the lesser kestrel, were largely confined to southern Spain's Magdalenian sites (e.g., the Cave of Erralla; Eastham 1985). In contrast, peregrine falcons, common kestrels, and Eurasian hobbies, which could tolerate both the colder temperatures and the subsequent warmer, drier conditions, were present in northern Spain and France (Fig. 2.3; Appendix; Eastham 1990). Gyrfalcons continued to inhabit continuous permafrost tundra regions in France (Fig. 2.3; Appendix). Additionally, adaptable species like the common kestrel reached Belgium during the Late Magdalenian, as indicated by the evidence of human exploitation (Goffette et al. 2020).

2.4.2 *Mesolithic (Western Europe) and Epipaleolithic (Levant)*

The Epipaleolithic period, which which spans from the last glacial maximum to the Holocene transition, was marked by significant climatic amelioration (Rosen and Rivera-Collazo 2012; Lloveras et al. 2020). Prior to the Neolithic, nomadic hunter-gatherers lived in small, seasonal camps rather than establishing permanent settlements (Shea 2013). The Natufian culture of the Levant exemplifies a sedentary, terminal Epipaleolithic hunter-gatherer society (Munro 2003). In European archaeology, the term "*Mesolithic*" is commonly used, while "*Epipaleolithic*" is preferred in Near Eastern archaeology (Fig. 2.4).

Fossil assemblages consist solely of leg (tarsometatarsi) and foot elements (phalanges) of falcons, which appear to have been selectively collected and transported to the home base (Tchernov 1993; Simmons and Nadel 1998; Valla et al. 2004). Evidence from numerous sites since the Mousterian period indicates that birds were hunted for various purposes, including obtaining feathers or skeletal parts for symbolic reasons, or as a food source (see Sect. 1.4; Finlayson et al. 2012; Martínez Valle et al. 2016; Lloveras et al. 2018). Although Tchernov (1993) proposed that sharp claws were used as tools, it is more likely that they served decorative or symbolic functions (Valla et al. 2004; Kuhn et al. 2004; Martin et al. 2013; Lloveras et al. 2020). While falcons were probably consumed by humans, they were only occasionally included in hunter-gatherer diets (Kuhn et al. 2004; Martin et al. 2013).

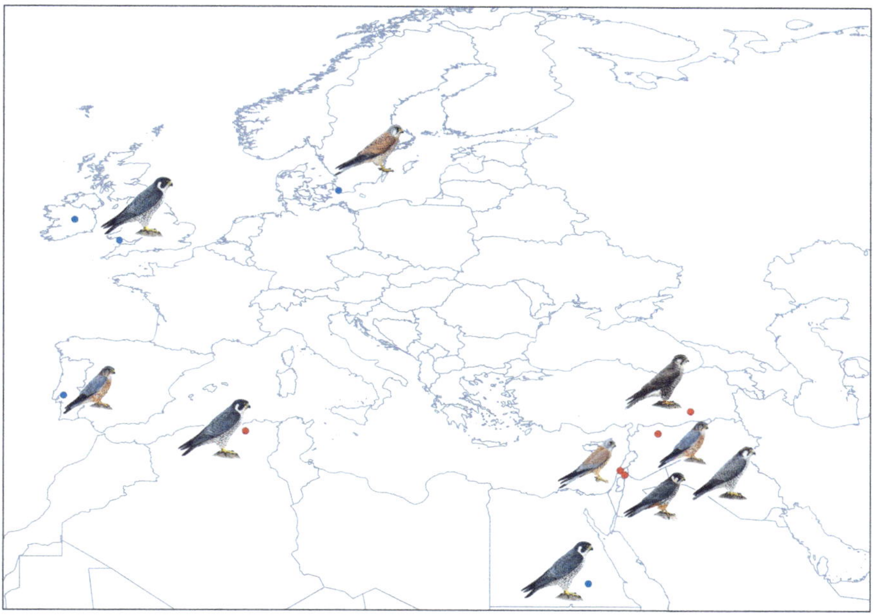

Fig. 2.4 Distribution of falcon fossils at archaeological sites from the Mesolithic (blue dots) and Epipaleolithic (red dots) industries. (Data from Appendix)

Hunting falcons is a difficult practice, especially when targeting adult birds, though juvenile or inexperienced ones might have been easier to capture (Kuhn et al. 2004). It is possible that these birds were not actively hunted; rather, feathers, claws, and talons were gathered from carcasses found in

the field (Yeomans and Ritcher 2018). Instruments discovered at a Final Natufian site (approximately 12,000 years ago) are capable of producing sounds similar to those of common kestrel (Davin et al. 2023). While these instruments may have been used as decoys to attract the falcons, the use of talons and feathers as personal ornaments suggests that mimicking falcon calls held significant symbolic value for these ancient communities (Davin et al. 2023).

During the Mesolithic/Epipaleolithic period, adaptable falcons were widespread, ranging from Egypt and the Maghreb to the United Kingdom (peregrine falcon) and from the Levant to southern Sweden (common kestrel; Fig. 2.4; Appendix; Tchernov 1968; Merzoug 2011; Spyrou et al. 2019). Saker falcons inhabited wooded steppe environments in Turkey (Fig. 2.4; Appendix; Zeder and Spitzer 2016). Additionally, lesser kestrels, common kestrels, Eurasian hobbies, merlins, peregrine falcons, and lanner falcons were present in the Levant (Fig. 2.4; Appendix; Zeder and Spitzer 2016). The opportunistic strategies employed by humans to capture or scavenge falcons in this region can offer valuable insights into the historical distribution and persistence of these predators (Table 2.2). The presence or absence of falcons in fossil assemblages may directly reflect their encounter rates in the field. Overall, there appears to be a decline in the number of species from the Late Natufian to the Neolithic periods, highlighting the persistence of more adaptable species (Table 2.2). Yeshurun et al. (2014) observed that the decline in both small and large falcons began during the Early Natufian (14,500–13,000 BP). During the Late Epipaleolithic period, warm and wet conditions prevailed, with the Levant covered by Mediterranean woodland/maquis, which was suitable for Eurasian hobbies and merlins (Langutt et al. 2021). Later, Mediterranean open field vegetation became dominant, providing an extensive herbaceous landscape that was ideal for habitat-generalist falcons such as the common kestrel (Langutt et al. 2021). Since the early Holocene, the region has been characterized by long, hot, dry summers and seasonal stress (Langutt et al. 2021).

Table 2.2 Falcon fossils from the Mureybet and Dja'de archaeological sites, spanning the Late Natufian through the Neolithic periods (including Khiamian, PPNA, and PPNB)

Species	Late Natufian (11,000–10,500 BP)	Khiamian (10,500–9800 BP)	PPNA (9800–8800 BP)	PPNB (8800–8300 BP)
F. tinnunculus	●	●	●	●
F. naumanni	●			
F. columbarius		●		
F. subbuteo	●	●		
F. peregrinus		●		●

PPNA pre-pottery neolithic A, *PPNB* pre-pottery neolithic B (Data from Gourichon and Helmer 2004)

2.4.3 *Neolithic Period*

The Neolithic period represents the final phase of the Stone Age. Although the significance of birds diminished with the rise of herding, they continued to play various roles in Neolithic communities (Russell 2018). Bird remains, including those of falcons, are generally rare in Neolithic sites across the Levant (see Table 2.2; Dayan et al. 1986). If the decline in game species is attributed to inadequate sieving or shifts in dietary habits, it is unlikely that falcons, primarily used for decorative purposes, would be prevalent (see Sect. 2.4.1). The remains reveal the diversity of falcons present in the Neolithic environment, a time when human hunting and scavenging activities were closely tied to areas of animal death (Best and Mulville 2018). Thus, falcons were likely captured or collected based on their availability in the field (see Sect. 2.4.2). Consequently, common and widespread species can be found across various pre-pottery Neolithic sites at different latitudes (e.g., common kestrel and peregrine falcon; Altuna et al. 2004; Peters et al. 2005; Best and Mulville 2018; Russell 2020; White et al. 2021a; Appendix). Conversely, local high abundances are expected to affect encounter rates in the field. For instance, although the lanner falcon is rare in both the Pleistocene and Holocene, it dominates the avian fossil assemblages in the Levant (see Chap. 1; Appendix; Simmons and Nadel 1998; White et al. 2021a). This interest in lanner falcons is supported by cut marks associated with wing or feather removal and the disarticulation of the tarsometatarsus from the tibiotarsus (Fig. 2.5; White et al. 2021b). Such interactions with falcons and butchering practices likely occurred across various latitudes, from the Levant to north-western Europe (i.e., Outer Hebrides; Best and Mulville 2018; Appendix). In the context of northern Neolithic hunters, falcons emerged as a significant element of the sun cult, arriving in the spring from the South to the subglacial tundra (Konstantinov et al. 2021).

2.5 Western Palearctic Falcons

All Western Palearctic falcon species persisted from the Pleistocene through the Holocene (see Sect. 1.6 and Fig. 2.6). However, some species and subspecies are believed to have become extinct during the Middle Pleistocene (Table 2.3; see also Sect. 1.4). Outside the Western Palearctic, the sole Holocene extinction was *Falco kurochkini*, a small falcon from Cuba (Suárez and Olson 2001; Tyrberg 2009).

A comprehensive study of fossils from the Carpathian Basin provides insights into the variation of falcon assemblages from the Late Miocene to the Holocene (Table 2.3; Kessler 2014). This extensive basin in south-eastern Central Europe was covered by open mesic grasslands from the Early/Late Pliocene to the Early Pleistocene (Szabó et al. 2022). The now-extinct common kestrel subspecies *F. t. atavus* inhabited the basin from the Late Miocene to the Middle Pleistocene (Table 2.3; see Sect. 1.4). By the early Pleistocene, a core assemblage of

Fig. 2.5 Tibiotarsus of a lanner falcon from the Neolithic WF16 site in southern Jordan, with cut marks indicated by arrows. (Credit: Steven Mithen)

open-habitat falcons had already emerged, including the red-footed falcon, merlin, Eurasian hobby, and saker falcon (Table 2.3). While extinct taxa such as *F. antiquus* and *F. t. atavus*, along with the gyrfalcon, disappeared between the Middle Pleistocene and the end of the Last Glacial Maximum, the core assemblage continued into the Holocene (Table 2.3). The transition from the Late Pleistocene to the Holocene impacted the distribution of extant falcon species but did not lead to their extinction. For example, species like the red-footed falcon persisted in the expanded forest-steppe lowlands, where dense grass cover was prevalent (Kovalchuk et al. 2021). Similarly, the peregrine falcon benefitted from the expansion of these open environments, which enlarged its wintering grounds (Nekrasov et al. 2016).

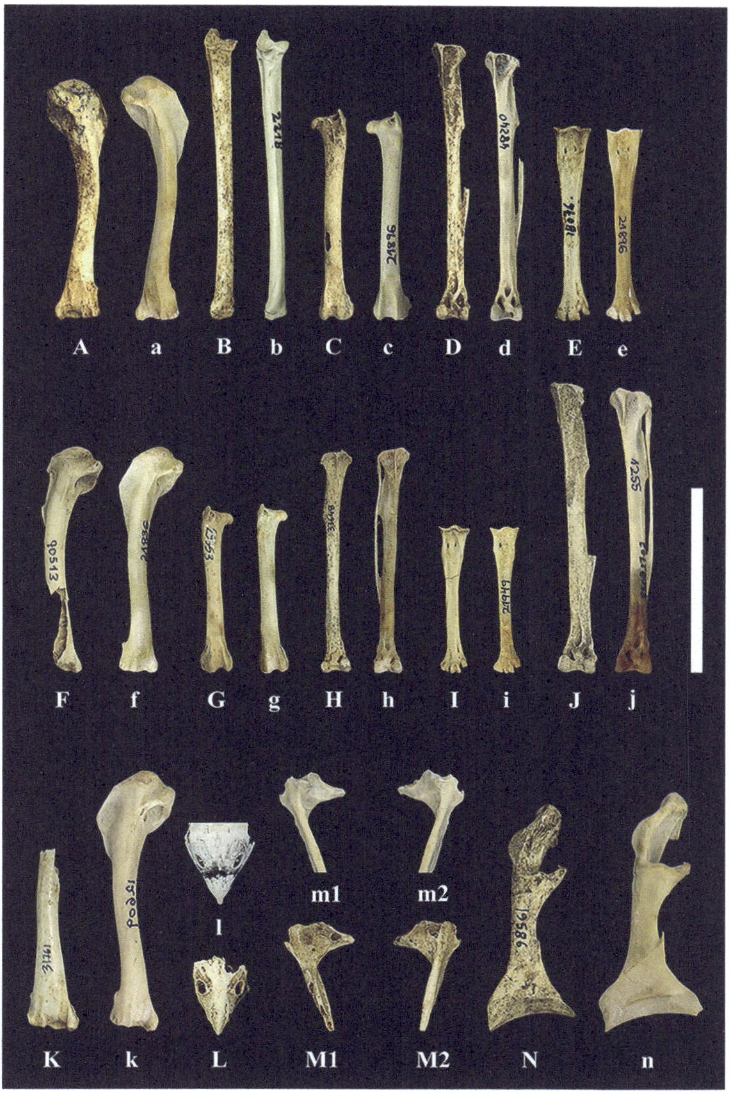

Fig. 2.6 *Falconidae* fossil specimens (capital letters) from Es Pouàs (Eivissa island, Pityusic archipelago, Western Mediterranean Sea) (lowercase letters: recent bones for comparison). A, a, B, b, C, c, D, d, E, e: *Falco tinunculus*. A: right humerus, cranial view; a: right humerus, cranial view; B: left ulna, ventral view; b: left ulna, ventral view; C: left femur, cranial view; c: left femur, cranial view; D: left tibiotarsus, cranial view; d: left tibiotarsus, cranial view; E: left tarsometatarsus, dorsal view; e: left tarsometatarsus, dorsal view. F, f, G, g, H, h, I, i: *Falco naumanni*. F: left humerus, cranial view; f: left humerus, cranial view; G: right femur, cranial view; g: right femur, cranial view; H: right tibiotarsus, cranial view; h: right tibiotarsus, cranial view; I: left tarsometatarsus, dorsal view; i: left tarsometatarsus, dorsal view. J, j: *Falco eleonorae*. J: left tibiotarsus, cranial view; j: left tibiotarsus, cranial view. K, k: *Falco subbuteo*. K: left humerus, cranial view; k: left humerus, cranial view. L: *Falco cf. subbuteo*, praemaxilla, dorsal view. l: *Falco subbuteo*, praemaxilla, dorsal view. M1, m1, M2, m2, N, n: *Falco peregrinus*. M1: mandible, dorsal view; m1: mandible, dorsal view; M2: mandible, ventral view; m2: mandible, ventral view; N: left coracoid, dorsal view; n: left coracoid, dorsal view. Scale bar: 4 cm. (Data from Guerra et al. 2013; Credit: Carmen Guerra)

Table 2.3 Fossilized remains of falcons from the Late Miocene to the Holocene have been discovered at various archaeological sites across the Carpathian Basin

Species	Late Miocene	Early Pliocene	Late Pliocene	Early Pleistocene	Middle Pleistocene	Late Pleistocene	Holocene
F. antiquus					●		
F. t atavus	●	●	●	●	●		
F. t. tinnunculus					●	●	●
F. vespertinus			●	●	●	●	●
F. columbarius				●	●	●	●
F. subbuteo				●	●	●	●
F. cherrug				●	●	●	●
F. rusticolus					●	●	
F. peregrinus						●	

Data from Kessler (2014)

However, in the latter part of the Late Holocene, the size of wintering sites diminished, and peregrine falcon migration routes shifted (Nekrasov et al. 2016).

The number of specimens in the falcon assemblage can be influenced by various factors, including the identity of fossil bone accumulators (see Sect. 2.4.2 for details on human activities). A substantial proportion of immature falcons can affect the total number of identified specimens (NISP), as evidenced by the presence of partially ossified bones (Fig. 2.7; Bochenski et al. 2018). In Sarakenos Cave, Greece, these immature falcons were likely preyed upon by eagle owls (*Bubo bubo*), which are adept at capturing inexperienced birds (Bochenski et al. 2018). The number of immature falcons increases markedly from the Paleolithic through the Mesolithic and Neolithic periods. In contrast, falcon subadult specimens were found only in the Paleolithic period (Bochenski et al. 2018). This pattern suggests that common kestrels and red-footed falcons were relatively abundant and accessible, with habitat conditions being more open rather than wooded (Fig. 2.7).

2.5.1 Falco naumanni

Lesser kestrels have been found to have survived in glacial refugia throughout the Holocene, including on the Iberian Peninsula and in southern Italy (Fig. 2.8). They were also present in the Levant, especially during the Late Epipaleolithic and Natufian periods (Appendix; Fig. 2.8).

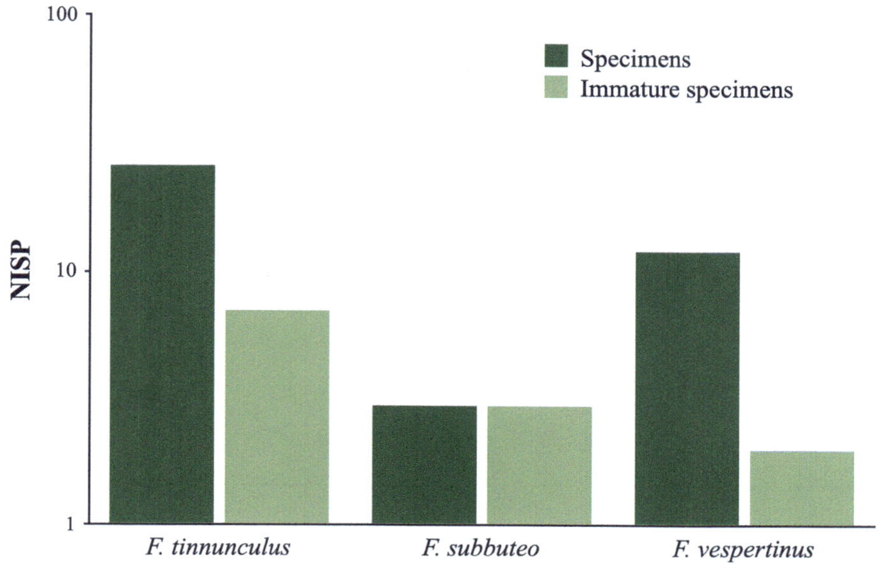

Fig. 2.7 Total and immature specimen counts (log-transformed) from Sarakenos Cave, Greece, during the Late Paleolithic (NISP = number of identified specimens; data from Bochenski et al. 2018)

2.5.2 Falco tinnunculus

The adaptable common kestrel was able to survive in various habitats, including discontinuous permafrost areas in France during the PHT (Castel et al. 2006; Appendix; Fig. 2.9). As the ice retreated and open habitats expanded, the common kestrel population also increased, as observed in England and southern Sweden (Yalden and Albarella 2009; Fig. 2.9). However, the spread of forests and the reduction of open areas hindered their colonization in Scandinavia (Breivik 2014).

2.5.3 Falco vespertinus

The red-footed falcon appears to have persisted in central eastern Europe since the Late Pliocene (Kessler 2014; Table 2.3). Despite westward expansion during the Pleistocene, this species remains relatively rare in PHT and Holocene archaeological sites (Fig. 2.10). With the exception of one site in Crimea, all these locations fall within the current migration pathways of the red-footed falcon (Fig. 2.10).

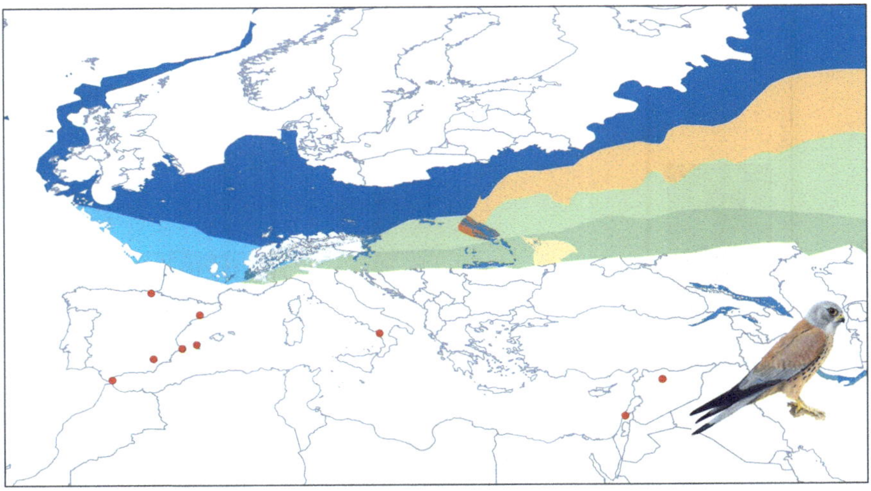

Fig. 2.8 Archaeological sites (red dots) with fossilized remains of lesser kestrel from the PHT and Holocene. Biome classifications for the Last Glacial Maximum permafrost regions are adapted from Lindgren (2018). Tundra—discontinuous permafrost = light blue; tundra—continuous permafrost = blue; steppe—discontinuous permafrost = green; steppe—continuous permafrost = light green. (Drawings courtesy of Marco Preziosi)

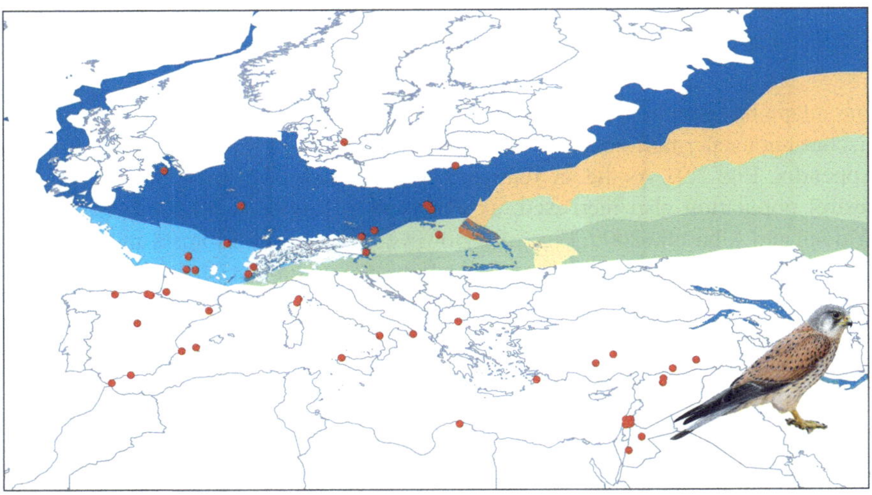

Fig. 2.9 Archaeological sites (red dots) with fossilized remains of common kestrel from the PHT and Holocene. Biome classifications for the Last Glacial Maximum permafrost regions are adapted from Lindgren (2018). Tundra—discontinuous permafrost = light blue; tundra—continuous permafrost = blue; steppe—discontinuous permafrost = green; steppe—continuous permafrost = light green. (Drawings courtesy of Marco Preziosi)

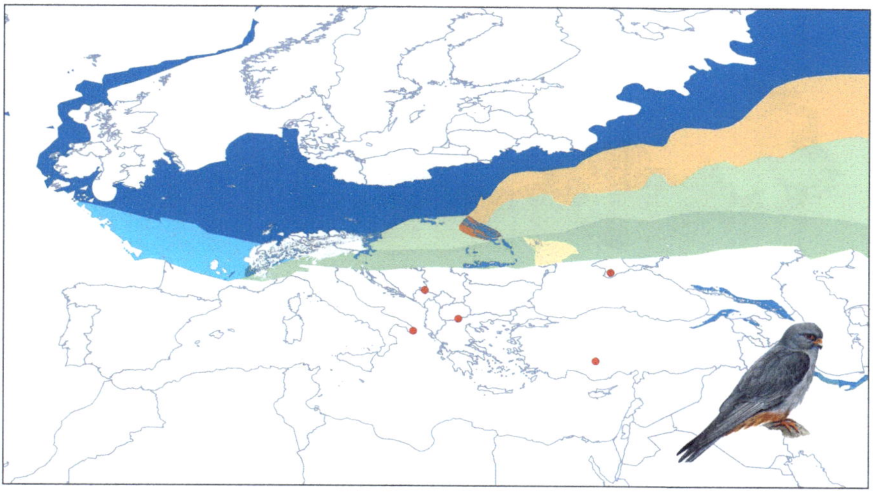

Fig. 2.10 Archaeological sites (red dots) with fossilized remains of red-footed falcon from the PHT and Holocene. Biome classifications for the Last Glacial Maximum permafrost regions are adapted from Lindgren (2018). Tundra—discontinuous permafrost = light blue; tundra—continuous permafrost = blue; steppe—discontinuous permafrost = green; steppe—continuous permafrost = light green. (Drawings courtesy of Marco Preziosi)

2.5.4 Falco eleonorae

Like the lesser kestrel, Eleonora's falcon continues to inhabit glacial refugia in the southern Mediterranean (Fig. 2.11). Their preference for the central western Mediterranean persisted throughout the PHT and Holocene, just as it did in the earlier Pleistocene epoch (Fig. 2.11; see also Sect. 1.6.4 and Fig. 1.15).

2.5.5 Falco columbarius

During Pleistocene cold periods, the merlin's preferred habitats shifted significant southward, allowing it to occupy territories in the Mediterranean (see Fig. 1.16). The species was present in the Levant and southern Europe during the PHT (Fig. 2.12). After deglaciation, the merlin moved to its current breeding areas in northern regions, previously covered by continuous permafrost and ice caps (Fig. 2.12).

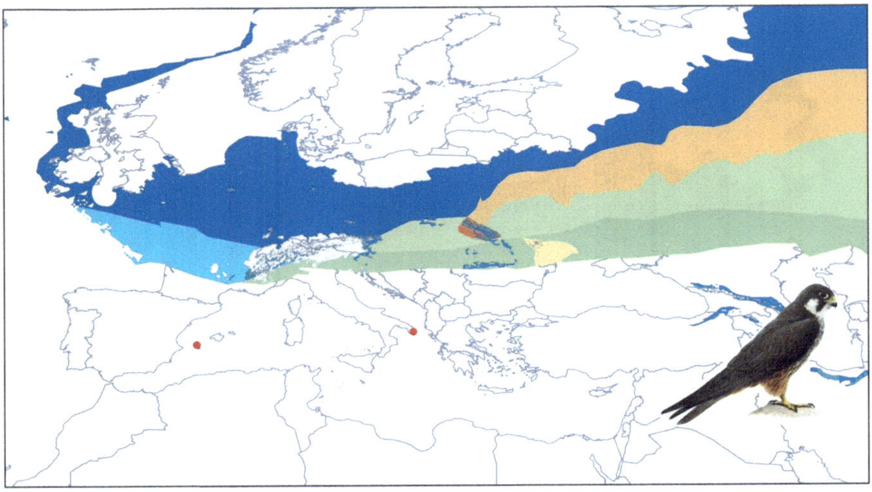

Fig. 2.11 Archaeological sites (red dots) with fossilized remains of Eleonora's falcon from the PHT and Holocene. Biome classifications for the Last Glacial Maximum permafrost regions are adapted from Lindgren (2018). Tundra—discontinuous permafrost = light blue; tundra—continuous permafrost = blue; steppe—discontinuous permafrost = green; steppe—continuous permafrost = light green. (Drawings courtesy of Marco Preziosi)

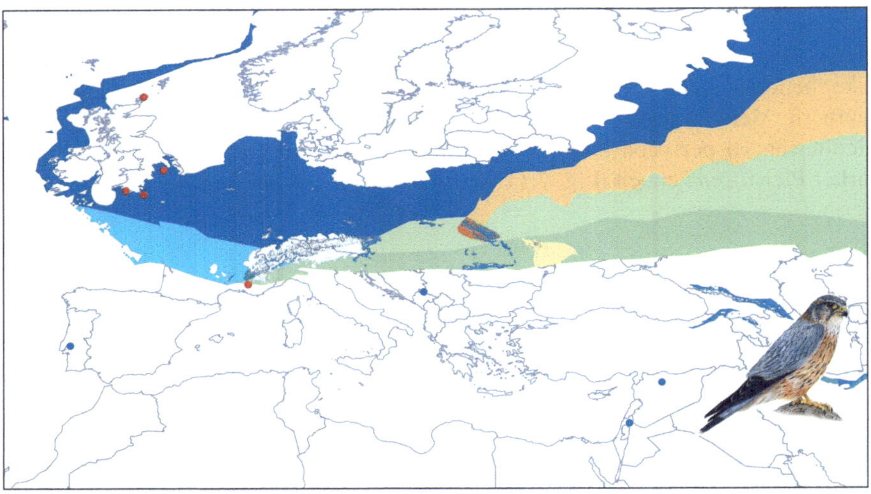

Fig. 2.12 Archaeological sites with fossilized remains of merlin from the PHT (blue dots) and Holocene (red dots). Biome classifications for the Last Glacial Maximum permafrost regions are adapted from Lindgren (2018). Tundra—discontinuous permafrost = light blue; tundra—continuous permafrost = blue; steppe—discontinuous permafrost = green; steppe—continuous permafrost = light green. (Drawings courtesy of Marco Preziosi)

2.5.6 Falco subbuteo

During the Holocene, the Eurasian hobby spread northward as permafrost retreated and forest ecotonal zones expanded (Fig. 2.13). Nevertheless, the species continued to persist in southern Europe, the Levant, and regions near discontinuous permafrost that it had occupied during the PHT (Fig. 2.13).

2.5.7 Falco biarmicus

The PHT and the Holocene, similar to the Late Pleistocene, are marked by a scarcity of fossil records for the lanner falcon (Fig. 2.14). Additionally, all the few fossils that have been found originate from the same Pleistocene region, the Levant (Fig. 2.14), leaving the complete absence of fossils unexplained in most of the lanner falcon's current range, including Italy, the Balkans, Turkey, and North Africa (Fig. 2.14; see also Fig. 1.18).

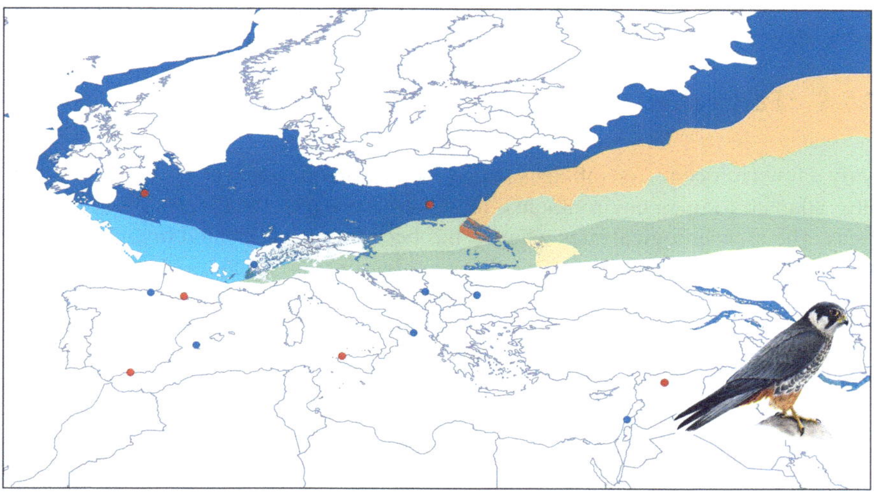

Fig. 2.13 Archaeological sites with fossilized remains of Eurasian hobby from the PHT (blue dots) and Holocene (red dots). Biome classifications for the Last Glacial Maximum permafrost regions are adapted from Lindgren (2018). Tundra—discontinuous permafrost = light blue; tundra—continuous permafrost = blue; steppe—discontinuous permafrost = green; steppe—continuous permafrost = light green. (Drawings courtesy of Marco Preziosi)

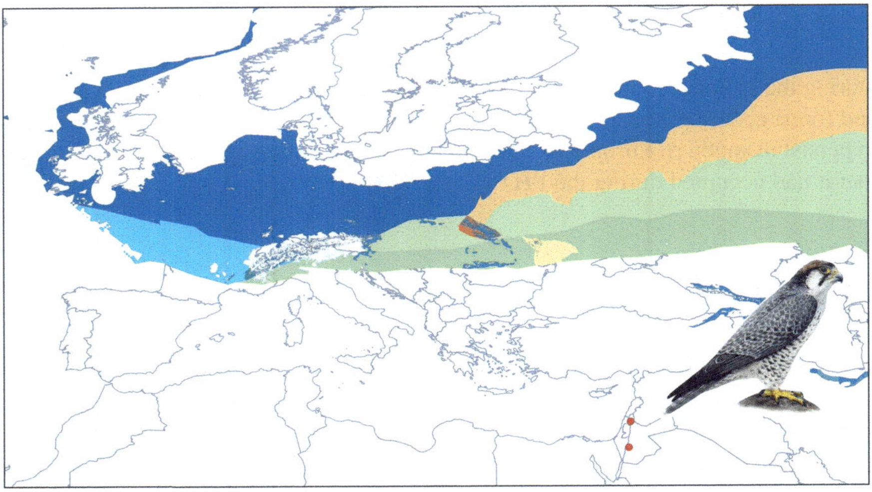

Fig. 2.14 Archaeological sites (red dots) with fossilized remains of lanner falcon from the PHT and Holocene. Biome classifications for the Last Glacial Maximum permafrost regions are adapted from Lindgren (2018). Tundra—discontinuous permafrost = light blue; tundra—continuous permafrost = blue; steppe—discontinuous permafrost = green; steppe—continuous permafrost = light green. (Drawings courtesy of Marco Preziosi)

2.5.8 Falco cherrug

Only a few Holocene assemblages include saker falcon fossils (Fig. 2.15). This species, adapted to open environments, likely suffered from the expansion of forested areas. The archaeological sites where it has been found are located in regions where it currently winters or, at most, breeds as a migrant (Fig. 2.15).

2.5.9 Falco rusticolus

The gyrfalcon has undergone a significant contraction in its distribution range, now confined to Arctic regions (Fig. 2.16). Archaeological sites where the species has been found correspond to areas of continuous and discontinuous permafrost during the PHT, which are distant from its breeding and wintering zones.

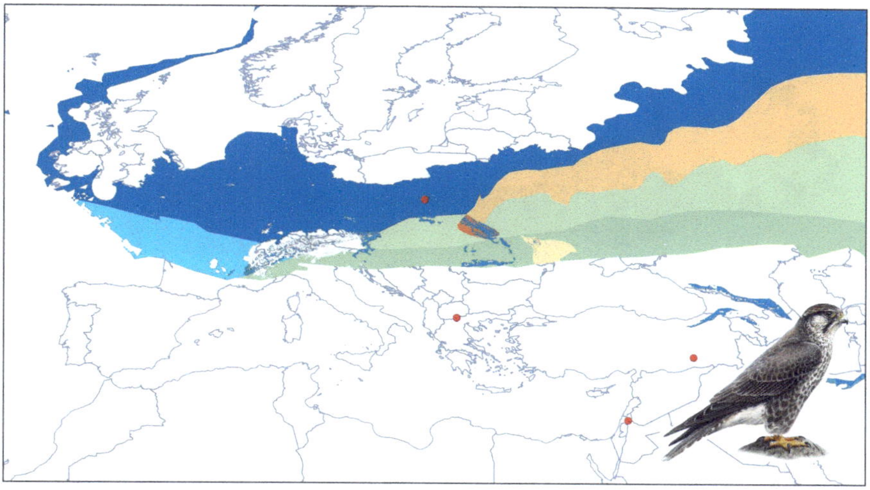

Fig. 2.15 Archaeological sites (red dots) with fossilized remains of saker falcon from the PHT and Holocene. Biome classifications for the Last Glacial Maximum permafrost regions are adapted from Lindgren (2018). Tundra—discontinuous permafrost = light blue; tundra—continuous permafrost = blue; steppe—discontinuous permafrost = green; steppe—continuous permafrost = light green. (Drawings courtesy of Marco Preziosi)

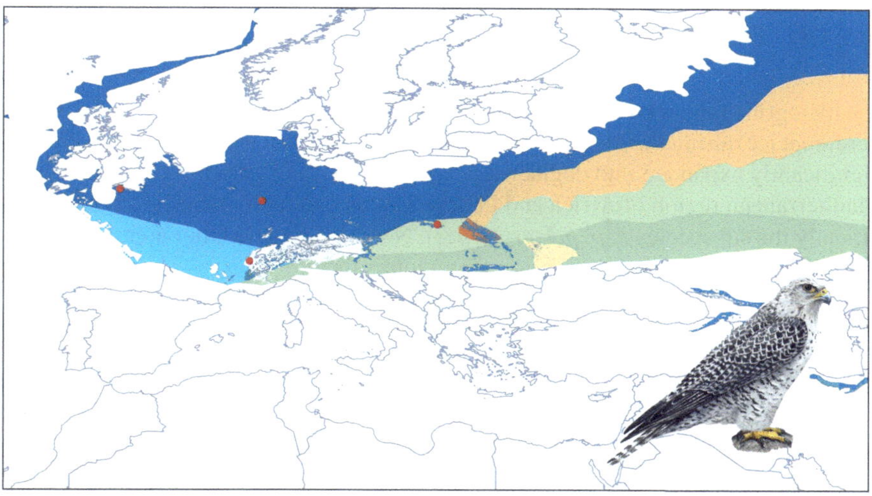

Fig. 2.16 Archaeological sites (red dots) with fossilized remains of gyrfalcon from the PHT and Holocene. Biome classifications for the Last Glacial Maximum permafrost regions are adapted from Lindgren (2018). Tundra—discontinuous permafrost = light blue; tundra—continuous permafrost = blue; steppe—discontinuous permafrost = green; steppe—continuous permafrost = light green. (Drawings courtesy of Marco Preziosi)

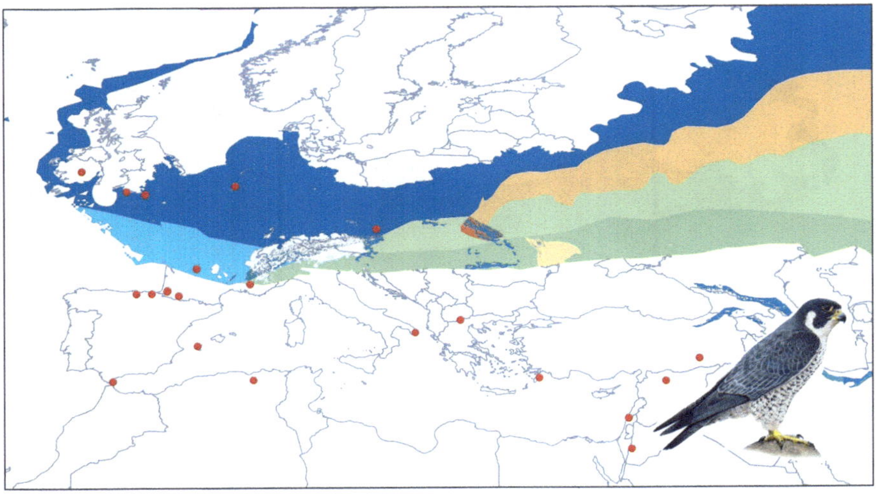

Fig. 2.17 Archaeological sites (red dots) with fossilized remains of peregrine falcon from the PHT and Holocene. Biome classifications for the Last Glacial Maximum permafrost regions are adapted from Lindgren (2018). Tundra—discontinuous permafrost = light blue; tundra—continuous permafrost = blue; steppe—discontinuous permafrost = green; steppe—continuous permafrost = light green. (Drawings courtesy of Marco Preziosi)

2.5.10 Falco peregrinus

The peregrine falcon has been observed migrating in the areas that were covered by perennial ice during the Pleistocene (Figs. 1.21 and 2.16). Over time, the species has significantly expanded into deglaciated regions (Fig. 2.17). Currently, it nests in the Mediterranean region of Africa and in desert areas, including the Middle East (specifically the subspecies *F. p. pelegrinoides*). Notably, in these latter areas, the nominal subspecies also has a population that winters there.

2.6 Conclusions

The Early Holocene was marked by rapid permafrost degradation and woodland expansion following the Pleistocene–Holocene transition. This period saw the replacement of Arctic and sub-Arctic species with more temperate fauna from glacial refugia. Although the transition from the Late Pleistocene to the Holocene impacted the distribution of extant falcon species, it did not lead to their extinction. Local variations occurred due to habitat changes, particularly with the expansion or retreat of open habitats as forests advanced.

Several key points can be highlighted:

- Warm-adapted falcons, such as the lesser kestrel and Eleonora's falcon, remained in glacial refugia in southern Europe.
- Boreal species, such as the merlin, occupied far northern open regions after deglaciation.
- Arctic species, such as the gyrfalcon, greatly reduced their range from permafrost-covered central Europe to their current breeding range (see Fig. 1.20).
- Temperate and adaptable species, such as the common kestrel, peregrine falcon, and Eurasian hobby, expanded into deglaciated areas.
- The expansion of forests hindered species of dry open habitats, such as the red-footed and saker falcons.
- The scarcity of lanner falcon fossils provide no insight into their movements during the PHT and Holocene (see also Sect. 1.6.7 for the Pleistocene).

Appendix: Fossil Remains of Western Palearctic Falcons from the Pleistocene–Holocene Transition (PHT) and the Holocene

Species	Dates	Location	References
F. naumanni	9850 ± 80 BP (Hol)	Cova de la Guineu, Spain	Galobart et al. (1991)
F. naumanni	12,500–11,000 BP (Natufian)	Hayonim, Israel	Tchernov (1993)
F. naumanni	12,500–11,000 BP (Natufian)	Mureybet, Syria	Tchernov (1993)
F. naumanni	12,660–12,950 BP (U Magdal)	Cova des Cendres, Spain	Villaverde et al. (1997)
F. naumanni	Late Magdalenian	Cave of Erralla, Spain	Eastham (1985)
F. naumanni	11,000–10,500 BP (Natufian)	Mureybet, Syria	Gourichon and Helmer (2004)
F. naumanni	16,420–18,440 BP (Solutrean)	Gorham's Cave, Gibraltar	Tyrberg (2008)
F. naumanni	16,500 ± 280 BP (Solutrean)	Cueva de Ambrosio, Spain	Tyrberg (2008)
F. naumanni	Upper Pleistocene and Holocene	Es Pouàs, Spain	Guerra et al. (2013)
F. naumanni	12,100–9020 BP (Epigravettian)	Grotta della Madonna, Italy	Gala et al. (2016)
F. naumanni	Early Neolithic	Göbekli Tepe, Turkey	Pöllath and Peters (2023)
F. tinnunculus	18,510 ± 580 BP (LGM/Hol)	Nixloch (Oberösterreich), Austria	Tyrberg (2008)
F. tinnunculus	9760 ± 160 BP (Hol)	Cingle Vermell, Spain	Vilette (1983)

(continued)

Species	Dates	Location	References
F. tinnunculus	11,700–10,500 BP (Early Hol)	Hallan Çemi, Turkey	Zeder and Spitzer (2016)
F. tinnunculus	Early–Middle Epipalaeolithic	Kharaneh IV, Jordan	Spyrou et al. (2019)
F. tinnunculus	7100–6000 BCE (Neolithic)	Çatalhöyük, Turkey	Russell (2018)
F. tinnunculus	6500–6000 BCE (Neolithic)	Pınarbaşı, Turkey	Russell (2020)
F. tinnunculus	8790–8840 BP (Neolithic)	Yiftah'el, Israel	Horwitz and Lernau (2003)
F. tinnunculus	ca. 14,000 (Magdalenian)	Cueva de Nerja, Spain	Cortés-Sánchez et al. (2008)
F. tinnunculus	8718–8300 BCE (Early Hol)	A Teppa di U Lupinu, Corse, France	Salotti et al. (2008)
F. tinnunculus	Pleist–Hol transition	Deszczowa, Poland	Cyrek et al. (2000)
F. tinnunculus	Early Holocene	Loutra Almopias Cave, Greece	Boev and Tsoukala (2019)
F. tinnunculus	12,500–11,000 BP (Natufian)	Hayonim, Israel	Tchernov (1993)
F. tinnunculus	12,500–11,000 BP (Natufian)	Mureybet, Syria	Tchernov (1993)
F. tinnunculus	Final Natufian	Ain Mallaha (Eynan), Israel	Valla et al. (2004)
F. tinnunculus	4200–3800 BCE (L Neolithic)	Dudka, Poland	Tomek and Gumiński (2003)
F. tinnunculus	17,000–12,000 (Magdalenian)	Cave of Amalda, Spain	Eastham (1990)
F. tinnunculus	12,660–12,950 BP (U Magdal)	Cova des Cendres, Spain	Villaverde et al. (1997)
F. tinnunculus	ca. 14,000 (Azilien)	Cueva de Santa Catalina, Spain	Laroulandie (2014)
F. tinnunculus	9000–8000 (Neolithic)	Göbekli Tepe, Turkey	Peters et al. (2005)
F. tinnunculus	11,800–10,200 BCE (Neolithic)	WF16 in Faynan, Jordan	White et al. (2021a)
F. tinnunculus	Younger Dryas	Gritulu Cave, Corse, France	Vigne et al. (1997)
F. tinnunculus	Early Neolithic	Asikli Hoyük, Turkey	Buitenhuis (1997)
F. tinnunculus	Late Magdalenian	Grotte des Freydières, France	Bouchud and Desbrosse (1973)
F. tinnunculus	Mesolithic	Grotta dell'Uzzo, Italy	Tagliacozzo (1994)
F. tinnunculus	Magdalenian	Abri du Lagopède, France	Schmider et al. (1995)

(continued)

Species	Dates	Location	References
F. tinnunculus	Epipaleolithic	Jerf el Ahmar, Syria	Gourichon (2002)
F. tinnunculus	9,640 ± 120 BP (Younger Dryas)	Pont d'Ambon, France	Jones (2009)
F. tinnunculus	ca.19,400 (Early Epipaleolithic)	Ohalo II, Israel	Simmons and Nadel (1998)
F. tinnunculus	Early Holocene	Biśnik Cave, Poland	Tomek et al. (2012)
F. tinnunculus	Neolithic	Kalythies Cave, Greece	Halstead and Jones (1987)
F. tinnunculus	11,800–9050 (Younger Dryas)	Grotta Romanelli, Italy	Tyrberg (2008)
F. tinnunculus	Magdalenian	Gorham's Cave, Gibraltar	Tyrberg (2008)
F. tinnunculus	Magdalenian	Gudenushöhle, Austria	Tyrberg (2008)
F. tinnunculus	13,690 BP (Mid-Late Würmian)	Grosse Ofenbergerhöhle, Austria	Tyrberg (2008)
F. tinnunculus	Holocene	Nixloch, Austria	Tyrberg (2008)
F. tinnunculus	12,370–12,990 BP (Magdalenian)	Trou de Chaleux, Belgium	Tyrberg (2008)
F. tinnunculus	13,600–13,920 BP (Allerød)	Temnata cave, Bulgaria	Tyrberg (2008)
F. tinnunculus	11,700–12,000 BP (Alleröd)	Abri Dufaure, France	Tyrberg (2008)
F. tinnunculus	11,030 BP (Late Magdalenian)	Bois-Ragot, France	Tyrberg (2008)
F. tinnunculus	13,000 BP (Magdalenian)	Grotte Jean-Pierre No. 1, France	Tyrberg (2008)
F. tinnunculus	14,890–19630 (L Solutrean)	Combe Saunière, France	Tyrberg (2008)
F. tinnunculus	Magdalenian	Vaskapu, Hungary	Tyrberg (2008)
F. tinnunculus	Upper Pleistocene and Holocene	Es Pouàs, Spain	Guerra et al. (2013)
F. tinnunculus	10,690–12,270 BP (Magdalenian)	La Riera, Spain	Tyrberg (2008)
F. tinnunculus	10,600–20,000 BP (Epipaleol.)	Haua Fteah, Lybia	Tyrberg (2008)
F. tinnunculus	ca. 13,500 BP	Tonnel'naya cave, Uzbekistan	Tyrberg (2008)
F. tinnunculus	Magdalenian	Jarama II, Spain	Tyrberg (2008)
F. tinnunculus	12,100–9020 BP (Epigravettian)	Grotta della Madonna. Italy	Gala et al. (2016)

(continued)

Species	Dates	Location	References
F. tinnunculus	Solutrean	Higueral-Guardia, Spain	Delgado Arcos (2019)
F. tinnunculus	Solutrean	Coves de Santa Maira, Spain	Tortosa et al. (2012)
F. tinnunculus	Mesolithic	Segebro, Sweden	Ericson and Tyrberg (2004)
F. tinnunculus	Late Devensian	Pinhole Cave, UK	Yalden and Albarella (2009)
F. tinnunculus	Late Paleolithic	Sarakenos Cave, Greece	Bochenski et al. (2018)
F. tinnunculus	Final Natufian	Eynan-Mallaha, Israel	Davin et al. (2023)
F. tinnunculus	Solutrean	Bois de Brousse, France	Eastham (2021)
F. tinnunculus	Solutrean	Arbreda, Spain	Eastham (2021)
F. tinnunculus	Solutrean	Arene Candide, Italy	Eastham (2021)
F. tinnunculus	12,000 BP (Late Natufian)	Nahal Ein Gev II, Israel	Munro et al. (2021)
F. vespertinus	7100–6000 BCE (Neolithic)	Çatalhöyük, Turkey	Russell (2018)
F. vespertinus	Early Holocene	Loutra Almopias Cave, Greece	Boev and Tsoukala (2019)
F. vespertinus	Bølling	Grot Skeliastyi, Ukraine	Kovalchuk et al. (2021)
F. vespertinus	10,390–9790 BP (Y Dryas)	Grotta Romanelli, Italy	Tyrberg (2008)
F. vespertinus	13,780 ± 140 BP	Malisina stijena, Montenegro	Tyrberg (2008)
F. vespertinus	Late Paleolithic	Sarakenos Cave, Greece	Bochenski et al. (2018)
F. eleonorae	10,390–9790 BP (Y Dryas)	Grotta Romanelli, Italy	Tyrberg (2008)
F. eleonorae	Upper Pleistocene and Holocene	Es Pouàs, Spain	Guerra et al. (2013)
F. eleonorae	Late Paleolithic	Sarakenos Cave, Greece	Bochenski et al. (2018)
F. eleonorae?	Magdalenian	Isturitz, France	Eastham (2021)
F. columbarius	Mesolithic	Cabeço da Amoreira, Portugal	Detry (2007)
F. columbarius	Early Neolithic	Göbekli Tepe, Turkey	Pöllath and Peters (2023)
F. columbarius	Neolithic	Aven Souche n#1, France	Crégut-Bonnoure et al. (2014)
F. columbarius	Neolithic	Calf of Eday, Scotland	Calder (1937)

(continued)

Species	Dates	Location	References
F. columbarius	10,500–9800 BCE (Khiamien)	Mureybet, Syria	Gourichon and Helmer (2004)
F. columbarius	ca.19,400 (Early Epipaleolithic)	Ohalo II, Israel	Simmons and Nadel (1998)
F. columbarius	13,780 ± 140 BP	Malisina stijena, Montenegro	Tyrberg (2008)
F. columbarius	Late Devensian	Pinhole Cave, UK	Yalden and Albarella (2009)
F. columbarius	Late Glacial	Soldier's Hole, UK	Yalden and Albarella (2009)
F. columbarius	Late Glacial	Robin Hood's Hole, UK	Yalden and Albarella (2009)
F. columbarius	Late Glacial	Cat Hole, UK	Yalden and Albarella (2009)
F. subbuteo	Early Holocene	Cueva de Nerja, Spain	Cortés-Sánchez et al. (2008)
F. subbuteo	12,500–11,000 BP (Natufian)	Hayonim, Israel	Tchernov (1993)
F. subbuteo	12,500–11,000 BP (Natufian)	Mureybet, Syria	Tchernov (1993)
F. subbuteo	Late Magdalenian	Cave of Erralla, Spain	Eastham (1985)
F. subbuteo	Magdalenian	Grotte du bois du Cantet, France	Clot and Omnès (1979)
F. subbuteo	Neolithic	Grotta dell'Uzzo, Italy	Tagliacozzo (1994)
F. subbuteo	10,500–9800 BCE (Khiamien)	Mureybet, Syria	Gourichon and Helmer (2004)
F. subbuteo	Early Holocene	Biśnik Cave, Poland	Tomek et al. (2012)
F. subbuteo	11,800–9050 (Younger Dryas)	Grotta Romanelli, Italy	Tyrberg (2008)
F. subbuteo	13,600–13,920 BP (Allerød)	Temnata cave, Bulgaria	Tyrberg (2008)
F. subbuteo	13,000 BP (Magdalenian)	Grotte Jean-Pierre No. 1, France	Tyrberg (2008)
F. subbuteo	Upper Pleistocene and Holocene	Es Pouàs, Spain	Guerra et al. (2013)
F. subbuteo	13,780 ± 140 BP	Malisina stijena, Montenegro	Tyrberg (2008)
F. subbuteo	12,100–9020 BP (Epigravettian)	Grotta della Madonna, Italy	Gala et al. (2016)
F. subbuteo	Late Glacial	Gough's Old Cave, UK	Yalden and Albarella (2009)
F. subbuteo	Late Paleolithic	Sarakenos Cave, Greece	Bochenski et al. (2018)
F. subbuteo	Magdalenian	Isturitz, France	Eastham (2021)
F. biarmicus	11,800–10,200 BCE (Neolithic)	WF16 in Faynan, Jordan	White et al. (2021a, b)
F. biarmicus	ca. 19,400 (Early Epipaleolithic)	Ohalo II, Israel	Simmons and Nadel (1998)

(continued)

Species	Dates	Location	References
F. cherrug	11,700–10,500 BP (L Epipal)	Hallan Çemi, Turkey	Zeder and Spitzer (2016)
F. cherrug	16,700–9900 BP (Late glacial)	Komarowa Cave, Poland	Tomek and Bochenski (2005)
F. cherrug	Early Holocene	Loutra Almopias Cave, Greece	Boev and Tsoukala (2019)
F. cherrug	ca.19,400 (Early Epipaleolithic)	Ohalo II, Israel	Simmons and Nadel (1998)
F. cherrug	ca. 13,500 BP	Tonnel'naya cave, Uzbekistan	Tyrberg (2008)
F. cherrug	Early Neolithic	Göbekli Tepe, Turkey	Pöllath and Peters (2023)
F. cherrug	12,000 BP (Late Natufian)	Nahal Ein Gev II, Israel	Munro et al. (2021)
F. rusticolus	12,980 ± 240 BP (Magdalenian)	Grotte des Romans, France	Desbrosse and Mourer-Chauviré (1973)
F. rusticolus	ca. 10,000 BP (Younger Dryas)	Karstein, Germany	Baales (1999)
F. rusticolus	ca. 12,400 BP (Older Dryas)	Jaskyňa v Hrdošnej skale Cave, SK	Obuch (2019)
F. rusticolus	Late Glacial	Caldey Island, UK	David (1991)
F. peregrinus	Early Neolithic	Göbekli Tepe, Turkey	Pöllath and Peters (2023)
F. peregrinus	11,700–10,500 BP	Hallan Çemi, Turkey	Zeder and Spitzer (2016)
F. peregrinus	3240 ± 40 BP (Middle Hol)	Toksanbai, Kazakhstan	Nekrasov et al. (2016)
F. peregrinus	9000 BCE (Mesolithic)	Aswan, Egypt	Tchernov (1968)
F. peregrinus	8000–4000 BCE (Mesolithic)	Lough Boora, Ireland	Warren (2015)
F. peregrinus	ca. 3000 BCE (Neolithic)	Schipluiden, The Netherlands	Zeiler (2006)
F. peregrinus	Early Neolithic	El Miron cave, Spain	Altuna et al. (2004)
F. peregrinus	Early Holocene	Loutra Almopias Cave, Greece	Boev and Tsoukala (2019)
F. peregrinus	Oldest Dryas	Coulet des Roches, France	Crégut-Bonnoure et al. (2014)
F. peregrinus	17,000–12,000 (Magdalenian)	Cave of Amalda, Spain	Eastham (1990)
F. peregrinus	11,800–10,200 BCE (Neolithic)	WF16 in Faynan, Jordan	White et al. (2021a)
F. peregrinus	10,207–8,416 (Epipaleolithic)	Medjez II, Algeria	Merzoug (2011)
F. peregrinus	10,500–9800 BCE (Khiamien)	Mureybet, Syria	Gourichon and Helmer (2004)
F. peregrinus	Neolithic	Kalythies Cave, Greece	Halstead and Jones (1987)
F. peregrinus	11,800–9050 (Younger Dryas)	Grotta Romanelli, Italy	Tyrberg (2008)

(continued)

Species	Dates	Location	References
F. peregrinus	10,700–12,400 BP (Natufian)	Hayonim cave, Israel	Tyrberg (2008)
F. peregrinus	Holocene	Gorham's Cave, Gibraltar	Tyrberg (2008)
F. peregrinus	Magdalenian	Gudenushöhle, Austria	Tyrberg (2008)
F. peregrinus	11,700–12,000 BP (Alleröd)	Abri Dufaure, France	Tyrberg (2008)
F. peregrinus	14,890–19,630 (L Solutrean)	Combe Saunière, France	Tyrberg (2008)
F. peregrinus	Upper Pleistocene and Holocene	Es Pouàs, Spain	Guerra et al. (2013)
F. peregrinus	Magdalenian	Grotte de Bourrouilla, France	Tyrberg (2008)
F. peregrinus	11,630 ± 280 BP	Gouffre du haut de la Coume, FR	Tyrberg (2008)
F. peregrinus	Late Glacial	Aveline's Hole, Somerset, UK	Yalden and Albarella (2009)
F. peregrinus	Late Glacial	Soldier's Hole, UK	Yalden and Albarella (2009)
F. peregrinus	Mesolithic	Port Eynon Cave, UK	Yalden and Albarella (2009)
F. peregrinus	Mesolithic	Gough's Cave, UK	Yalden and Albarella (2009)
F. peregrinus	Late Paleolithic	Sarakenos Cave, Greece	Bochenski et al. (2018)
F. peregrinus	Magdalenian	Cueva de Nerja, Spain	Eastham (2021)
Falco sp.	Mousterian	Les Fieux, France	–

References

Altuna J, Cuenca Bescós G, Elorza M, García Pimienta JC, Mariezkurrena K et al (2004) Post-Pleistocene faunas from the Archeological site of El Miron Cave (Ramales de la Victoria, Cantabria, Spain): A preliminary summary. In: Baquedano E, Rubio Jara S (eds) Miscelánea en homenaje a Emiliano Aguirre. Zona Arqueológica, Alcalá de Henares. II: Paleontologia, Zona Arqueológica 4. Museo Arqueológico Regional, Alcalá de Henares, pp 40–49

Aura Tortosa JE, Morales Pérez JV, Carrión Marco Y, Avezuela Aristo B et al (2012) Coves de Santa Maira (Castell de Castells, La Marina Alta, Alacant). First data on the Solutrean occupation. In: Avezuela Aristu B, Jordá Pardo JF (eds) Congreso Internacional el Solutrense. Spain, pp 61–62

Baales M (1999) Economy and seasonality in the Ahrensburgian. In: Kozlowski SK, Gruba J, Zaliznyak LL (eds) Tanged points cultures in Europe (colloquium Lublin 1993). Lubelskie materialy archeologiczne 13, pp 64–75

Banks WE, d'Errico F, Peterson AT, Vanhaeren M, Kageyama M, Sepulchre P, Ramstein G, Jost A, Lunt D (2008) Human ecological niches and ranges during the LGM in Europe derived from an application of eco-cultural niche modeling. J Archaeol Sci 35:481–491. https://doi.org/10.1016/j.jas.2007.05.011

Best J, Mulville J (2018) Birds in death: avian archaeology and the mortuary record in the Scottish Islands. In: Livarda A, Madgwick M, Riera S (eds) Bioarchaeology of ritual and religion. Oxbow, Oxford, pp 179–192

Bocheński ZM, Tomek T, Wertz K, Kaczanowska M, Kozłowski JK, Sampson A (2018) Who ate the birds: the taphonomy of Sarakenos Cave, Greece. Archaeol Anthropol Sci 10:1603–1615. https://doi.org/10.1007/s12520-017-0488-3

Boev Z, Tsoukala E (2019) Late Pleistocene and Earliest Holocene avifauna from the Loutra Almopias Cave (Macedonia, Greece). Historia naturalis bulgarica 40:1–31

Bortolini E, Pagani L, Oxilia G, Posth C, Fontana F, Badino F et al (2021) Early Alpine occupation backdates westward human migration in Late Glacial Europe. Curr Biol 31:1–10. https://doi.org/10.1016/j.cub.2021.03.078

Bouchud J, Desbrosse R (1973) La faune de la Grotte des Freydières à Saint-Agnan- en Vercors (Drôme). Bull Société Préhistorique Française 70:330–334

Breivik HM (2014) Palaeo-oceanographic development and human adaptive strategies in the Pleistocene–Holocene transition: a study from the Norwegian coast. The Holocene 24:1478–1490. https://doi.org/10.1177/0959683614544061

Buitenhuis H (1997) Aşikli Höyük: a "protodomestication" site. Anthropozoologica 25–26:655–662

Calder CST (1937) A Neolithic double-chambered cairn on the Calf of Eday, Orkney. Proc Soc Antiqu Scotl 71:115–154

Castel J-C, Liolios D, Laroulandie V, Chauvière F-X, Chadelle J-P, Pike-Tay A, Geneste J-M (2006) Solutrean animal resource exploitation at Combe Saunière (Dordogne, France). In: Maltby M (ed) Integrating Zooarchaeology, Durham, pp 138–172

Clot A, Omnès J (1979) Premiers datages radiocarbone du Magdalénien des Hautes-Pyrénées. Bull Société Préhistorique Française 76:324–339

Cortés Sánchez M, Morales-Muñiz A, Simón-Vallejo MD, Bergadà-Zapata MM et al (2008) Palaeoenvironmental and cultural dynamics of the coast of Málaga (Andalusia, Spain) during the Upper Pleistocene and early Holocene. Quart Sci Rev 27:2176–2193. https://doi.org/10.1016/j.quascirev.2008.03.010

Crees JJ, Collen B, Turvey ST (2019) Bias, incompleteness and the 'known unknowns' in the Holocene faunal record. Philos Trans R Soc B 374:20190216. https://doi.org/10.1098/rstb.2019.0216

Crégut-Bonnoure E, Argant J, Bailon S et al (2014) The karst of the Vaucluse, an exceptional record for the Last Glacial Maximum (LGM) and the Late-glacial period palaeoenvironment of south-eastern France. Quat Int 339–340:41–61. https://doi.org/10.1016/j.quaint.2013.10.058

Cyrek K, Nadachowski A, Madeyska T et al (2000) Excavations in the Deszczowa Cave (Kroczyckie rocks, Czestochowa Upland, Central Poland). Folia Quaternaria 71:5–84

David A (1991) Late Glacial archaeological residues from Wales: a selection. In: Barton N, Roberts AJ, Roe DA (eds) The late glacial in north-west Europe: human adaptation and environmental change at the end of the Pleistocene. CBA research report 77, pp 141–159

Davin L, Tejero J-M, Simmons T, Shaham D, Borvon A, Tourny O, Bridault A et al (2023) Bone aerophones from Eynan-Mallaha (Israel) indicate imitation of raptor calls by the last hunter-gatherers in the Levant. Sci Rep 13:8709. https://doi.org/10.1038/s41598-023-35700-9

Davis BAS, Brewer S, Stevenson AC, Guiot J (2003) The temperature of Europe during the Holocene reconstructed from pollen data. Quat Sci Rev 22:1701–1716. https://doi.org/10.1016/S0277-3791(03)00173-2

Dayan T, Tchernov E, Bar-Yosef O, Yom-Tov Y (1986) Animal exploitation in Ujrat el Mehed, a Neolithic site in southern Sinai. Paléorient 12:105–116

Delgado Arcos JA (2019) Zooarcheological and taphonomical study of the faunal bone remains of the Mousterian and Solutrean levels of the Higueral-Guardia cave (Malaga-Cadiz). Inferences in subsistence strategies and palaeoecology. Arqueologia y Territorio 16:1–22

Desbrosse R, Mourer-Chauviré C (1973) Les oiseaux magdaleniens de Pierre-Chatel (Ain). Quartärely 23(24):149–164

Detry C (2007) Paleoecologia e paleoeconomia do Baixo Tejo no Mesolítico final: o contributo do estudo dos mamíferos dos concheiros de Muge. Unpublished Ph.D. Dissertation. Universidade Autónoma de Lisboa, Lisboa/Salamanca

Driver JC, Hobson KA (1992) A 10 500-year sequence of bird remains from the southern boreal forest region of western Canada. Arctic 45:105–110. https://doi.org/10.14430/arctic1379

Eastham A (1985) The magdalenian avifauna at Erralla cave. In: Altuna J et al (eds) Cazadores magdalenienses en Erralla. Munibe 37, San Sebastián, pp 59–80

Eastham A (1990) The bird bones in the cave of Amalda. In: La cueva de Amalda (Zestoa, País Vasco): ocupaciones paleolíticas y postpaleolíticas. Colección Barandiaran, Eusko Ikaskuntz, pp 239–253

Eastham A (2021) Man and bird in the palaeolithic of western Europe. Archaeopress Publishing, Oxford

Enloe J (2001) Magdalenian. In: Peregrine PN, Ember M (eds) Encyclopedia of Prehistory, Volume 4: Europe. Springer, New York. https://doi.org/10.1007/978-1-4615-1187-8

Ericson PGE, Tyrberg T (2004) The early history of the Swedish avifauna: a review of the subfossil record and early written sources. Kungliga Vitterhets Historie och Antikvitets Akademiens Handlingar, Antikvariska Serien 45:1–349

Finlayson C, Brown K, Blasco R, Rosell J, Negro JJ et al (2012) Birds of a feather: neanderthal exploitation of raptors and corvids. PLoS ONE 7(9):e45927. https://doi.org/10.1371/journal.pone.0045927

Gala M, Fiore I, Tagliacozzo A (2016) The raptors in the Mesolithic at the Grotta del Santuario della Madonna at Praia a Mare (Cosenza, Italy). Preistoria Alp 48:215–229

Galobart A, Garcia L, Güell A, Millán M, Ros MT, Serrano G (1991) Estudi de la fauna i flora fòssils de la Cova de la Guineu i el seu entorn. Technical report. Generalitat de Catalunya, Barcelona

Goffette Q, Germonpré M, Lefèvre C et al (2020) Bird bones from Trou de Chaleux and the human exploitation of birds during the late Magdalenian in Belgium. J Archaeol Sci Rep 29:102096. https://doi.org/10.1016/j.jasrep.2019.102096

Gourichon L (2002) Bird remains from Jerf el Ahmar, a PPNA site in Northern Syria, with special reference to the griffon vulture (*Gyps fulvus*). In: Buitenhuis H, Choyke AM, Mashkour M, Al-Shiyab AH (eds) Archaeozoology of the near East. Center for Archeological Research and Consultancy, Rijksuniversiteit Groningen, Groningen, pp 139–153

Gourichon L, Helmer D (2004) Exploitation de petit gibier dans le Moyen Euphrate syrien du XIIe au IXe millénaires av. J.-C. In: Petits animaux et sociétés humaines. Du complement alimentaire aux ressources utilitaires. Éditions APDCA, pp 415–433

Guerra C, McMinn M, Alcover JA (2013) The Upper Pleistocene-Holocene raptorial bird guild from Eivissa Island (Pityusic Archipelago, Western Mediterranean Sea). Geobios 46:491–502. https://doi.org/10.1016/j.geobios.2013.10.003

Halstead P, Jones G (1987) Bioarchaeological remains from Kalythies cave, Rhodes. In: Sampson A (ed) I Neolithiki Periodos sta Dodekanisa. Ministry of Culture, Athens, pp 135–152

Hewitt GM (1999) Post-glacial re-colonization of European biota. Biol J Linn Soc 68:87–112. https://doi.org/10.1111/j.1095-8312.1999.tb01160.x

Horwitz LK, Lernau O (2003) Temporal and spatial variation in Neolithic caprine exploitation strategies: a case study of fauna from the site of Yiftah'el (Israel). Paléorient 29(1):19–58

Jones EL (2009) Climate change, patch choice, and intensification at Pont d'Ambon (Dordogne, France) during the Younger Dryas. Quat Res 72:371–376. https://doi.org/10.1016/j.yqres.2009.08.003

Kalis AJ, Merkt J, Wunderlich J (2003) Environmental changes during the Holocene climatic optimum in central Europe—human impact and natural causes. Quat Sci Rev 22:33–79. https://doi.org/10.1016/S0277-3791(02)00181-6

Kessler J (2014) Fossil and subfossil bird remains and faunas from the Carpathian Basin. Ornis Hungarica 22:65–125. https://doi.org/10.2478/orhu-2014-0019

Konstantinov AV, Pimenov NV, Pavlova AV, Ivannikova RF (2021) Features of the microbiota of Falconiformes birds. IOP Conf Ser Earth Environ Sci 677:042043

Kovalchuk O, Rekovets L, Tsvelykh A, Yanenko V, Manko V, Tajkova S (2021) Living in a time of change: late pleistocene/holocene transitional vertebrate fauna of Grot Skeliastyi (Crimea, Ukraine). Hist Biol 33. https://doi.org/10.1080/08912963.2020.1769094

Kuhn SL, Belfer-Cohen A, Barzilai O, Stiner MC, Kerry KW, Munro ND, Bar-Yosef Mayer D (2004) The Last Glacial Maximum at Meged rockshelter, Upper Galilee, Israel. J Israel Prehist Soc 34:5–47

Lambeck K, Purcell A, Zhao J, Svensson N (2010) The Scandinavian ice sheet: from MIS 4 to the end of the Last Glacial Maximum. Boreas 39:410–435. https://doi.org/10.1111/j.1502-3885.2010.00140.x

Langgut D, Cheddadi R, Sharon G (2021) Climate and environmental reconstruction of the Epipaleolithic Mediterranean Levant (22.0-11.9 ka cal. BP). Quat Sci Rev 270:107170. https://doi.org/10.1016/j.quascirev.2021.107170

Laroulandie V (2000) Taphonomie et archéozoologie des oiseaux en grotte: applications aux sites Paléolithiques du Bois-Ragot (Vienne), de Combe Saunière (Dordogne) et de La Vache (Ariège). PhD thesis, Université de Bordeaux, Bordeaux

Laroulandie V (2014) Traitement et utilisation des ressources aviaires au Tardiglaciaire dans la Grotte de Santa Catalina, Kobie. Excavaciones Arqueológicas en Bizkaia 4:297–330

Lindgren A (2018) Biomes of the Last Glacial Maximum permafrost region. Dataset version 1. Bolin Centre Database. https://doi.org/10.17043/lindgren-2018-biomes-1

Lloveras L, Garcia L, Maroto J, Soler J, Soler N (2018) The bird assemblage from the Middle Palaeolithic level I of Arbreda Cave: a taphonomic story. J Archaeol Sci Rep 21:758–770. https://doi.org/10.1016/j.jasrep.2018.08.040

Lloveras L, Salazar R, García-Argüelles P, Nadal J (2020) Birds and Epipalaeolithic hunter-gatherers in northeast Iberia. The case of the Balma del Gai site. Quat Int 543:25–33. https://doi.org/10.1016/j.quaint.2020.03.010

Martin L, Edwards Y, Garrard A (2013) Broad spectrum or specialised activity? Birds and tortoises at the Epipalaeolithic site of Wadi Jilat 22 in the eastern Jordan steppe. Antiquity 87:649–665. https://doi.org/10.1017/S0003598X00049371

Martínez Valle R, Guillem Calatayud PM, Villaverde Bonilla V (2016) Bird consumption in the final stage of Cova Negra (Xátiva, Valencia). Quat Int 421:85–102. https://doi.org/10.1016/j.quaint.2016.01.068

Merzoug S (2011) Faunal remains from Medjez II (Epipaleolithic, Algeria): evidence of ostrich consumption and interpretation of Capsian subsistence behaviors. In: Jousse H, Lesur J (eds) People and animals in holocene Africa: recent advances in archaeozoology. Africa Magna Verlag, Frankfurt am Main, pp 123–131

Mol J (2008) Definition of the time slices. Landscape and climate change during the Last Glaciation in Europe; a review. In: Markova A, van Kolfschoten T (eds) Evolution of European ecosystems during pleistocene e holocene transition (24-8 kyr BP). KMK Scientific Press, Moscow, pp 73–87

Munro N (2003) Small game, the Younger Dryas, and the transition to agriculture in the southern Levant. Mitteilungen der Gesellschaft für Urgeschichte 12:47–71

Munro ND, Petrillo AN, Grosman L (2021) Specialized aquatic resource exploitation at the Late Natufian site of Nahal Ein Gev II, Israel. Archaeol Anthropol Sci 13:6. https://doi.org/10.1007/s12520-020-01257-1

Naughton F, Sanchez Goñi MF, Rodrigues T, Salgueiro E, Costas S, Desprat S, Duprat J, Michel E, Rossignol L, Zaragosi S, Abrantes F (2016) Climate variability across the last deglaciation in NW Iberia and its margin. Quat Int 414:9–22. https://doi.org/10.1016/j.quaint.2015.08.073

Nekrasov AE, Kosintsev PA, Samashev Z et al (2016) New data on avifauna of the Ustyurt plateau in the Holocene. Dokl Biol Sci 469(1):170–172

Obuch J (2019) Birds and small mammals dated to the end of Pleistocene found in the caves of Veľká Fatra Mts. Slov Kras acta carsol Slov 57:75–81

Peters J, von den Driesch A, Pöllath N, Schmidt K (2005) Birds in the megalithic art of Pre-Pottery Neolithic Göbekli Tepe, Southeast Turkey. In: Grupe G, Peters J (eds) Feathers, grit and symbolism. Birds and humans in the ancient old and new worlds. Proceedings of the 5th meeting of the ICAZ Bird Working Group in Munich, Rahden/Westf. Verlag Marie Leidorf, pp 223–234

Pöllath N, Peters J (2023) Distinct modes and intensity of bird exploitation at the dawn of agriculture in the Upper Euphrates and Tigris River basins. Archaeol Anthropol Sci 15:154. https://doi.org/10.1007/s12520-023-01841-1

Renssen H, Seppä H, Heiri O, Roche DM, Goosse H, Fichefet T (2009) The spatial and temporal complexity of the Holocene thermal maximum. Nat Geosci 2:411–414. https://doi.org/10.1038/ngeo513

Rognon P (1987) Late Quaternary climatic reconstruction for the Maghreb (North Africa). Palaeogeogr Palaeoclimatol Palaeoecol 58:11–34. https://doi.org/10.1016/0031-0182(87)90003-4

Rosen A, Rivera-Collazo I (2012) Climate change, adaptive cycles, and persistence of foraging economies during the late Pleistocene/Holocene transition in the Levant. PNAS USA 109:3640–3645. https://doi.org/10.1073/pnas.1113931109

Russell N (2018) Feathers and talons: birds at Neolithic Çatalhöyük, Turkey. Archaeol Anthropol Sci 11:6393–6410. https://doi.org/10.1007/s12520-018-0681-z

Russell N (2020) Changing use of birds across the agricultural transition at Pınarbaşı, Turkey. Quat Int 626–627:43–51. https://doi.org/10.1016/j.quaint.2020.10.077

Salotti M, Louchart A, Bailon S, Lorenzo S, Oberlin C et al (2008) A Teppa di U Lupinu Cave (Corsica, France)—human presence since 8500 years BC, and the enigmatic origin of the earlier, late Pleistocene accumulation. Acta Zool Cracov 51A:15–34. https://doi.org/10.3409/azc.51a_1-2.15-34

Schmider B, Valentin B, Baffier D, David F, Julien M, Leroi-Gourhan A, Mourer-Chauviré C, Poulain T, Roblin-Jouve A, Taborin Y (1995) L'abri du Lagopède (fouilles Leroi-Gourhan) et le Magdalénien des grottes de la Cure (Yonne). Gallia Préhistoire 37:55–114

Shea JJ (2013) Stone tools in the Paleolithic and Neolithic Near East: a guide. Cambridge University Press, Cambridge

Simmons T, Nadel D (1998) The avifauna of the early Epi-palaeolithic site of Ohalo II (19–400 years BP), Israel: species diversity, habitat and seasonality. Int J Osteoarchaeol 8:79–96. https://doi.org/10.1002/(SICI)1099-1212(199803/04)8:2<79::AID-OA386>3.0.CO;2-I

Spyrou A, Maher LA, Martin LA, Macdonald DA, Garrard A (2019) Meat outside the freezer: Drying, smoking, salting and sealing meat in fat at an Epipalaeolithic megasite in eastern Jordan. J Anthropol Archaeol 54:84–101. https://doi.org/10.1016/j.jaa.2019.02.004

Stewart JR, Lister AM, Barnes I, Dalén L (2010) Refugia revisited: individualistic responses of species in space and time. Proc R Soc B Biol Sci 277:661–671. https://doi.org/10.1098/rspb.2009.1272

Stivrins N, Soininen J, Amon L, Fontana SL, Gryguc G, Heikkilä M, Heiri O, Kisielienė D, Reitalu T, Stančikaitė M, Veski S, Seppä H (2016) Biotic turnover rates during the Pleistocene-Holocene transition. Quat Sci Rev 151:100–110. https://doi.org/10.1016/j.quascirev.2016.09.008

Straus LG (2015) The human occupation of southwestern Europe during the last glacial maximum. Solutrean cultural adaptations in France and Iberia. J Anthropol Res 71:465e492. https://doi.org/10.3998/jar.0521004.0071.401

Suárez W, Olson SL (2001) A remarkable new species of small falcon from the Quaternary of Cuba (*Aves*: *Falconidae*: *Falco*). Proc Biol Soc Wash 114:34–41

Szabó P, Kovács J, Kocsis L, Vennemann T, Domingo L et al (2022) Pliocene—early Pleistocene continental climate and vegetation in Europe based on stable isotope compositions of mammal tooth enamel. Quat Sci Rev 288:107572. https://doi.org/10.1016/j.quascirev.2022.107572

Tagliacozzo A (1994) Economic changes between the Mesolithic and the Neolithic in the Grotta dell'Uzzo (Sicily, Italy). Accordia Res Papers 5:7–37

Tchernov E (1968) Peregrine Falcon and Purple Gallinule of late Pleistocene age in the Sudanese Aswan Reservoir area. Auk 85:133

Tchernov E (1993) Exploitation of birds during the Natufian and early Neolithic of the Southern Levant. Archaeofauna 2:121–143

Tomek T, Bochenski ZM (2005) Weichselian and Holocene bird remains from Komarowa Cave, Central Poland. Acta Zool Cracov 48A(1–2):43–65

Tomek T, Gumiński W (2003) Bird remains from the Mesolithic and Neolithic Site Dudka, Masuria, Poland. Acta Zool Cracov 46:9–18

Tomek T, Bocheński ZM, Socha P, Stefaniak K (2012) Continuous 300,000-year fossil record: changes in the ornithofauna of Biśnik Cave, Poland. Palaeontol Electron 15:2A–20p

Turvey ST, Blackburn TM (2011) Determinants of species abundance in the Quaternary vertebrate fossil record. Paleobiology 37:537–546. https://doi.org/10.1666/09075.1

Tyrberg T (2008) Pleistocene birds of the Palearctic. http://web.telia.com/~u11502098/pleistocene.pdf

Tyrberg T (2009) Holocene avian extinction. In: Turvey ST (ed) Holocene extinctions. Oxford University Press, Oxford, pp 63–106

Valla FR, Khalaily H, Valladas H, Tisnerat-Laborde N, Samuelian N, Bocquentin F, Rabinovich R, Bridault A, Le Dosseur G, Rosen A, Dubreuil L, Bar-Yosef D, Belfer-Cohen A (2004) Les fouilles de Mallaha en 2000 et 2001: 3eme rapport preliminaire. J Isr Prehist Soc 34:49–184

van Vliet-Lanoë V, Pissart A, Baize S, Brulhet J, Ego F (2019) Evidence of multiple thermokarst events in northeastern France and southern Belgium during the two last glaciations. A discussion on “Features caused by ground ice growth and decay in Late Pleistocene fluvial deposits, Paris basin, France” (Bertran et al., 2018). Geomorphology 327:613–628. https://doi.org/10.1016/j.geomorph.2018.08.036

Vigne J-D, Bailon S, Cuisin J (1997) Biostratigraphy of amphibians, reptiles, birds and mammals in Corsica and the role of man in the Holocene faunal turnover. Anthropozoologica 25–26:587–604

Vilette P (1983) Avifaunes du Pléistocène final et de l’Holocène dans le Sud de la France et en Catalogne. Laboratoire de Préhistoire et Palethnologie. Dépôt de Fouilles Préhistoriques, Carcassonne

Villaverde V, Martínez Valle R, Guillem PM, Badal E, Zalbidea L, García R (1997) Els nivells Magdalenians de la cova de les Cendres (Teulada, Moraira). Resultats del sondeig del quadre A-17. Aguaits 13/14:77–115

von den Driesch A, Pöllath N (2010) Distribution history of European wild birds based on archaeozoological records. In: Grupe G, McGlynn G, Peters J (eds) Archaeobiodiversity. A European perspective. Marie Leidorf, pp 71–216

Warren G (2015) ‘Mere food gatherers they, parasites upon nature…’: Food and drink in the Mesolithic of Ireland. Proc R Ir Acad 115C:1–26. https://doi.org/10.1353/ria.2015.0004

White J, Finlayson B, Makarewicz C et al (2021a) The bird remains from WF16, an early Neolithic settlement in southern Jordan: assemblage composition, chronology, and spatial distribution. Int J Osteoarchaeol 31:1030–1045. https://doi.org/10.1002/oa.3016

White J, Khoury F, Greet B et al (2021b) The utilization of birds at Neolithic WF16, southern Jordan: cut marks, body parts and experimental skinning. Int J Osteoarchaeol 31:1203–1216. https://doi.org/10.1002/oa.3031

Yalden DW, Albarella U (2009) The history of British birds. Oxford University Press, Oxford

Yeomans L, Ritcher T (2018) Exploitation of a seasonal resource: bird hunting during the Late Natufian at Shubayqa 1. Int J Osteoarchaeol 28:95–108. https://doi.org/10.1002/oa.2533

Yeshurun R, Bar-Oz G, Weinstein-Evron M (2014) Intensification and sedentism in the Terminal Pleistocene Natufian sequence of el-Wad Terrace (Israel). J Hum Evol 70:16–35. https://doi.org/10.1016/j.jhevol.2014.02.011

Zeder MA, Spitzer M (2016) New insights into the broad-spectrum revolution in the Near East: the birds of Hallan Çemi. Quat Sci Rev 151:140–159. https://doi.org/10.1016/j.quascirev.2016.08.024

Zeiler JT (2006) Birds. In: Louwe Kooijmans LP, Jongste PFB (eds) Schipluiden. A Neolithic settlement on the Dutch North Sea coast c. 3500 cal BC. Analecta Praehistorica Leidensia 37/38, Leiden, pp 421–442

Chapter 3
Ancient History and Late Antiquity

3.1 Introduction

This chapter introduces a shift in methodology from the preceding two chapters, which focused on the Pleistocene and Holocene periods (Chaps. 1 and 2). This adjustment is necessitated by a significant reduction in the number of fossils, particularly subfossils, compared to earlier epochs (see Appendix). In contrast, falcons have gained cultural prominence and are extensively featured in various art forms, including paintings, steles, and jewelry (see Appendix). They have also been revered, mummified, and elevated to a position of prestige and nobility, as seen in instances such as god transfiguration and falconry (see Box 3.1). However, it is important to note that these cultural representations predominantly apply to larger falcon species, while other species receive minimal or no historical mention. Consequently, certain species are disproportionately represented, while others are entirely absent from cultural records. Despite these limitations, data on the capture of individual falcons can still provide valuable insights. Falcons do not readily reproduce in captivity and, therefore, had to be captured in the wild, making capture data a significant source of information (Allsen 2006).

A persistent challenge lies in accurately associating various names and definitions of falcon species with their contemporary systematic classifications and ensuring the equivalence of names across different languages (see Box 3.2). This challenge is further complicated by the potential importation of exotic species for falconry, which adds to the confusion (Yalden and Albarella 2009; see Box 3.3). Similar issues arise in figurative arts, where falcons were not depicted with the precision found in ornithological manuals until the 16th century. Instead, their depictions were influenced by symbolic roles. Consequently, numerous efforts have been made to retroactively identify species mentioned in historical contexts across different periods. For example, ornithologists have endeavored to determine which falcon

G. Leonardi, *The History of Western Palearctic Falcons*,
https://doi.org/10.1007/978-3-031-86528-2_3

species corresponds to the personification of the Egyptian deity Horus (Wilkinson 1878; see Box 3.1).

Since the latter half of the twentieth century, this complex context has created a division between modern ornithology and the study of ancient history and literature. Therefore, when no fossil records are available, it is imperative to exercise extreme caution when interpreting documents and depictions. Reconstructing historical distribution solely through the inference from past or literary sources poses significant challenges in population ecology. A more favorable approach involves a mixed methodology: using fossil evidence as the primary foundation while incorporating historical observations whenever it is possible to accurately identify the species.

This chapter spans 4300 years, from the Bronze Age (fourth millennium BCE) through the first three centuries of the first millennium, marking the end of the era known as "*Ancient History.*" The following three centuries (fourth to sixth century CE) constitute the phase referred to as "*Late Antiquity.*" While this division into historical periods may not directly relate to the ecological aspects of the species, it helps contextualize the interplay between humans and these species.

3.2 Fourth Millennium BCE

The fourth millennium BCE witnessed significant cultural transformations, including the rise of the Bronze Age and the advent of written language. Notably, the Kingdom of Egypt and the city states of Sumer gained prominence, while agriculture spread across Eurasia.

In Egypt, the latter part of the fourth millennium is known as the Late Predynastic Period, which preceded Dynasty I (Campagno 2013). Recent discoveries of rock carvings at Gebel Tjauti (3200–3000 BCE) have revealed scenes depicting early symbols of kingship, including falcons (Darnell 2002). The renowned Narmer Palette illustrates the unification of Upper and Lower Egypt under King Narmer, with a falcon symbolizing Horus perched above a cluster of papyrus flowers, representing Lower Egypt.

In northern Europe, the remains of Eurasian hobby and common kestrel were found at Ajvide, a Pitted Ware culture site on the Island of Gotland in the Baltic Sea (Mannermaa and Storå 2006). These specimens appear to have been placed in the soil shortly after the birds' deaths or otherwise protected from environmental deterioration at this hunter-gatherer site, dating to approximately 3300 BCE (Mannermaa and Storå 2006). Additionally, at another archaeological site in Sweden, dating back to the Middle Neolithic period, the remnants of a peregrine falcon were discovered within a tomb (Ericson and Tyrberg 2004).

3.3 Third Millennium BCE

The third millennium BCE, known as the Early to Middle Bronze Age, witnessed the emergence of early empires in the Ancient Near East. The use and significance of birds in ancient Greece, particularly in the Bronze Age cultures around the Aegean Sea, have attracted scholarly interest (Binnberg 2018; Mylona 2022). Depictions of falcons on vase paintings and frescoes have been found in the Cyclades, Crete, and the Greek mainland, although discoveries of falcon bones are rare (Binnberg 2018; Mylona 2022). Falcons, possibly common kestrels and Eleonora's falcons, have been identified by their color, shape, and behavior (Binnberg 2018). A miniature fresco from Akrotiri portrays what is presumed to be an Eleonora's falcon as part of a local naval procession or festival, while another fresco from Phylakopi on Melos (2250–2000 BCE) depicts the narrow wings of two birds, possibly falcons (Binnberg 2018). However, bone data show the presence of ubiquitous species such as peregrine falcons and common kestrels (see Sect. 3.4; Mylona 2022).

Fossils of widespread species have been found in various locations, including the common kestrel (Tell Sweyhat, Syria), the peregrine falcon (Tofts Ness, Scotland), and the Eurasian hobby (Velikent, Daghestan; Appendix; Buitenhuis 1985; Morales-Muñiz and Antipina 2003; Serjeantson 2007). A noteworthy study examines the accumulation of feces in gyrfalcon nests in central-west and northwest Greenland (Burham et al. 2009). The oldest guano sample in the Kangerlussuaq region dates to around 2740–2360 BCE, while other nests date from 1480 to 1080 BCE (Burham et al. 2009). Remarkably, about 80% of the sites used by gyrfalcons during Ancient History and Late Antiquity are the same as those used during the Contemporary Period (twentieth to twenty-first century; Burham et al. 2009).

3.4 Second Millennium BCE

During the second millennium BCE, a significant transition occurred in the Ancient Near East from the Middle to the Late Bronze Age. The Middle Kingdom of Egypt and Babylonia dominated the early part of the millennium, followed by the rise of the Mycenaean Greeks in the Aegean and the ascendance of the Hittite Empire (Foxhall 1995; Murname 1995; George 2007; Bryce 2012).

At the Bronze Age archaeological site of Kommos in Crete, archaeologists discovered bones from common kestrels and peregrine falcons (Mylona 2022). Common kestrels have also been found at Tell el-Dab'a (Egypt, Second Intermediate Period–Early New Kingdom; ca. 1800–1500 BCE) and are depicted in mythological scenes where the goddesses Isis and Nephthys assume their appearance (Boessneck 1976; Houlinan and Goodman 1986). In the Dynasty XIX tomb of Queen Nefertari (QV66) in the Valley of the Queens (thirteenth century BCE), Nephthys is depicted as a reddish common kestrel (Houlinan and Goodman 1986). The king was symbolized by a falcon as the living Horus (see Box 3.1). During the

Box 3.1: Horus Worship

Since ancient Predynastic times, falcons have been considered manifestations of Horus, the sky god, or of Ra, the sun god (Fig. 3.1; Porter 2011; Ikram and Spitzer 2021). Egyptian art frequently depicts falcons carrying the sun in their wings across a blue arch, symbolizing the sky or heavens (Porter 2011). The standing falcon symbolized the connection between humans and gods, while the diving falcon represented death and functioned as a mortuary deity, specifically Sokar, the protector of the primary Necropolis in Memphis (Porter 2011). From the late Dynasty I period until the Ptolemaic/Roman era, falcon depictions remained largely unchanged, with only minor variations (Porter 2011). Artifacts emphasized symbolic and ornamental qualities rather than resembling actual wild birds. Certain color combinations, such as green tails, blue heads and necks, and blue spots on green wings, held significant magical meaning (Podgórski 2010). The common kestrel appears in Sennedjem's and Nefertari's tomb paintings, but the beautifully painted stone inside the Mortuary Temple Hatshepsut at Deir el-Bahari may not be attributable to a single species such as the peregrine falcon (Houlinan and Goodman 1986; Porter 2011). Nevertheless, the language of the ancient Egyptians indicates that they were aware of the transition in plumage color that occurred as falcons matured (Kenning 2002).

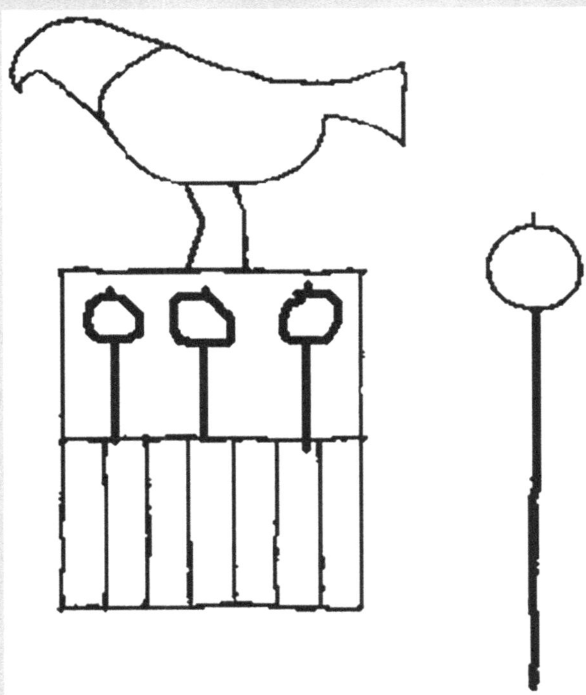

Fig. 3.1 Clay cutting of Hedju-Hor hieroglyph. (ca 3250 BCE; Credit: Wikipedia; https://en.wikipedia.org/wiki/Hedju_Hor#/media/File:Hedju.png)

Ramesside period (1292–1069 BCE), the king was often compared to a hunting falcon, which becomes furious when it sees small birds (Edgerton and Wilson 1936).

Other records from the second millennium include the discovery of *Falco* sp. bones at Saruq al-Hadid in the United Arab Emirates and a common kestrel at the great Hittite rock-cut sanctuary of Yazıkaya in Turkey (Appendix; Boessneck 1975; Roberts et al. 2020).

Between 660 BCE and the Roman era (around 340 CE), people began offering mummified animals, along with statues, statuettes, and stelae, to gods in hopes of having their prayers answered (Fig. 3.2; Allsen 2006; Ikram and Spitzer 2021). Consequently, falcon breeding became a commercial enterprise in Egypt, with birds being bred for sale (Warburton 2018; Ikram and Spitzer 2021). Although capturing and keeping raptors can be challenging, it is possible with some *Falco* species. Evidence suggests that a breeding program once existed to supply the mummy industry with small falcons, particularly common kestrels, which are frequently found (Ikram and Spitzer 2021). The common kestrel is the only falcon species that can be bred easily in captivity, and evidence from mummified specimens indicates that they were subjected to forced feeding, a common practice in breeding programs (Ikram et al. 2015). However, it is also possible that *juvenile* falcons were obtained through nest captures instead of captive breeding, as it is easier to catch nestlings and fledglings than adult birds.

Although vast quantities of mummified birds are entombed in underground catacombs in Egypt, only a small fraction has been thoroughly investigated (Bibb and McKnight 2022). Nonetheless, a significant number of mummified falcons from temples and animal cemeteries have been analyzed within fossilized assemblages (von den Driesch et al. 2006; Osypinska et al. 2020; Ikram and Spitzer 2021; see Fig. 3.3). The preserved animals in these tombs are predominantly birds (>99% *Aves*: 32% falcons, 27% ibises; Ikram and Spitzer 2021). Analyzing these collections of falcon fossils can provide valuable insights into the species that were present during that time and their distribution in the field (Fig. 3.3).

3.5 First Millennium BCE

The first millennium BCE encompasses the Iron Age and marks the transition from the early civilizations of the Ancient Near East, located in the Middle East and the Levant, to classical antiquity centered around Mediterranean Sea (Sherratt and Sherratt 1993; Manning 2018).

Several bone findings of small falcons from this time period have been made, including Eurasian hobbies in Sweden, common kestrels in Jordan and the Orkney Islands, and merlins in the same location (see Appendix; Ericson and Tyrberg 2004; Popkin 2009; Yalden and Albarella 2009). Among large falcons, noteworthy discoveries include *Falco cf. cherrug* in Sudan (late New Kingdom/pre-Napatan to Napatan times; 12–4 BCE) and a peregrine falcon in Siberia, where a more favorable, warmer climate prevailed during the early Iron Age (Appendix; Potapova and Panteleyev

Fig. 3.2 The figure, dating from the Late Period to the Ptolemaic Period (664–630 BCE), was part of a mummy box that contained a falcon mummy intended for placement in an animal necropolis. These boxes were linked not only to the worship of the prominent falcon god Horus but also with the wider Ra solar cult. (Met's open access images; https://www.metmuseum.org/art/collection/search/570747)

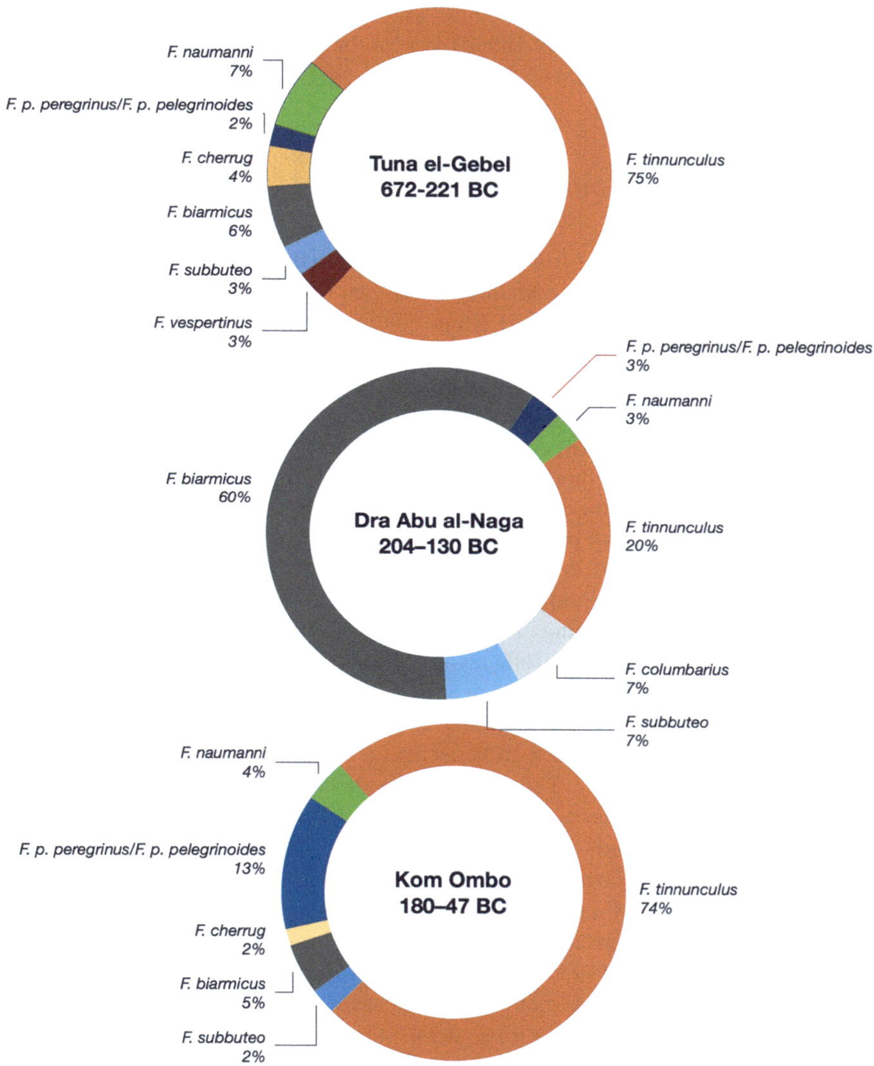

Fig. 3.3 Frequencies, presented as percentages, of mummified falcon species in three different subterranean animal necropolises in Egypt. (Data: von den Driesch et al. 2006; Bailleul-LeSuer 2019; Ikram and Spitzer 2021)

1999; Linseele and Pöllath 2015). Additionally, a gyrfalcon bone was found among 60,000 bird bones in Nipisat (Greenland), part of the Saqqaq Palaeo-Eskimo culture (Gotfredsen and Møbjerg 2004).

3.5.1 Classical Antiquity

Classical antiquity, spanning from the eighth century BCE to the fifth century CE, is renowned for its focus on the Mediterranean Sea region and includes the famous civilizations of ancient Greece and Rome, collectively known as the Hellenistic Roman world.

Notably, the ancient Greeks did not engage in falconry as a recreational pursuit (Barringer 2001). Aristotle referred to a bird of prey as "*kirkos*" (κίρκος), a term that Homer in the Odyssey (fourth century BCE) designated as "*Apollo's nimble messenger*" (Arnott 2007). Authors such as Aeschylus and Apollonius described this bird as a long-winged raptor capable of swift, direct flight, soaring to great heights before descending upon its prey—typically pigeons, but also starlings, jackdaws, hooded crows, ducks, and domestic fowl (Arnott 2007). While this description aligns well with the peregrine falcon, it is important to note that ancient terms like "*kirkos*" held multiple meanings and were often used broadly to denote a "*small bird of prey*" (see Box 3.2). Aristotle also mentions another type of falcon or hawk, calling it "*hupotriorkhes*" (literally near-buzzard), traditionally identified as the Eurasian hobby (*subbuteo* = near-buzzard; Arnott 2007). The Greek philosopher and naturalist Democritus described kestrels at dovecotes, likely referring to lesser kestrels (Negro et al. 2020). The construction of the Parthenon in Greece likely provided nesting sites for lesser kestrel colonies, which continued to inhabit the site until at least the nineteenth century (Brehm 1858). Aristophanes clearly describes the lesser kestrel in his comedy "*Birds*" (414 BCE), noting these small falcons that, in flocks, catch larger Orthoptera (i.e., locusts; Arnott 2007).

Falcon remains from archaeological sites dating to the Roman period are rare (Rielly 1985; Hernández 1993; Kroll, 2012; see Appendix). These remains primarily consist of smaller falcons, such as lesser kestrels in Spain and common kestrels in Roman Britain, discovered in forts, towns, and villas (Parker 1988; Hernández 1993). Notably, the distal half of the right femur of an adult common kestrel was found among domestic waste (Parker 1988). Additionally, the discovery of a Eurasian hobby at Settefinestre in Italy, dating back to the Roman Late Republic, suggests their possible use as decoys (Rielly 1985). Evidence indicate that the Romans did not practice falconry but rather used these birds for decoy purposes (Yalden and Albarella 2009). This is exemplified in a hunting scene depicted in a fourth-century mosaic from Sicily, where a man is shown with a falcon (Wilson 1983).

The ancient Egyptians assigned religious significance to specific bird species in grave goods and votive offerings (Morgan and McGovern-Huffman 2008; see also Box 3.1). According to Herodotus, pharaohs took stringent measures to protect these revered birds, even instituting the death penalty for those who mistreated or harmed them (von den Driesch et al. 2006). Various types of jars, heavily soaked in turpentine, were used to contain the bones of individual birds, alongside mummies crafted in the likeness of falcon-headed deities (Fig. 3.2; von den Driesch et al. 2006; Ikram and Spitzer 2021). Significant raptor assemblages have been discovered in the subterranean animal necropolises of the Saitic (672–525 BCE) and Ptolemaic (304–30 BCE) periods, in locations such as Tuna el-Gebel in Middle Egypt (672–221 BCE), Dra Abu al Naga west of Luxor (204–130 BCE), and the Kom Ombo Temple (180–47 BCE; see Fig. 3.3 and Appendix). Generally, common kestrels were the most frequently mummified falcon species, followed by lanner, peregrine, and saker falcons (Fig. 3.3; Houlinan and Meininger 1986; Gautier 2005). The high prevalence of common kestrels is unsurprising given their abundance and widespread distribution; however, it is also crucial to consider the captive breeding programs established by the Egyptians (see Fig. 3.3 and Box 3.1; Ikram et al. 2015). The unusually high frequency of lanner falcons (>60%) at the Dra Abu al Naga site (see Fig. 3.3) poses an interpretative challenge. If not attributable to captive breeding, this less common species might have experienced significant nest depredation. It is possible that, similar to Mesopotamia, local Egyptian populations claimed hereditary rights to specific nesting sites (Allsen 2006). Additionally, some mummified birds of prey, including falcons, were migratory species from outside the Nile Valley, as indicated by their bone and feather isotopic compositions (Linglin et al. 2020).

3.6 First Millennium CE

The first millennium of the Common Era (CE) spans from the year 1 to 1000 CE. This paragraph focuses on the initial three centuries (first to third c.), followed by the period from the fourth to the sixth century, known as Late Antiquity (see Sect. 3.6.1). The final centuries of this millennium, from the seventh to the tenth century, are classified as the Early Middle Ages (see Sect. 4.2). During the Roman Iron Age, the climate was relatively warm and dry, but around 400 CE, a sudden climatic shift occurred toward wetter summers and colder winters (Lucy 2001; Table 3.1). The Roman Warm Period (RWP, 1–250 CE) was a regional phenomenon characterized by notably warm weather in Europe and the North Atlantic (Wang et al. 2012; Shi et al. 2022). This period had observable effects on avian populations in certain regions, including a decline in cold and temperate species and an increase in

Table 3.1 Chronological framework of the Roman Iron Age, highlighting the Migration Period within the context of Late Antiquity (see Sect. 3.6.1)

Dates	Early Iron Age	Historical period
400–1 BCE		Pre-Roman Iron Age
1–400 CE		Roman Iron Age
400–550 CE		Migration period

temperate and subtropical species (Gorobets and Rudenko 2022). However, due to an insufficient dataset, a comprehensive analysis of these impacts on falcons is not feasible at present (Appendix). Nevertheless, as demonstrated by the gyrfalcons in Greenland (see Sect. 3.3), localized climate improvements can enhance nest site occupancy and increase the abundance of preferred prey (Burham et al. 2009; Moseley et al. 2019).

Fossils from this historical period are sparse, with common kestrels in Bulgaria being a notable example (Boev and Beech 2007). Another significant find is the barbary falcon discovered within a pottery sarcophagus at the pet cemetery in Berenice, Egypt, a Red Sea port (Osypinska et al. 2020; Appendix). Information about falcons in Roman literature is also limited. Aelian, in his work "*De natura animalium*," recounts a story about falconry in Thrace, originally reported by Aristotle and later cited by Pliny. Aelian noted that, in that region, falcons were used instead of hawks (Gualtieri 2005). The "*Physiologus*," considered the precursor to the bestiary genre, was likely written in Greek in Alexandria around the third or fourth century CE. However, it does not mention falcons or hawks at all (Gualtieri 2005).

3.6.1 Late Antiquity

Late Antiquity, which spanned from the fourth to the sixth century, marked the transition from classical antiquity to the Middle Ages in Europe and the Mediterranean basin. This period began with the decline and fall of the Western Roman Empire and concluded with the Muslim conquests of North Africa and Iberia (Cameron 2012). The latter part of Late Antiquity experienced a significant climatic shift known as the Late Antique Little Ice Age (LALIA; 536–660; Büntgen et al. 2016). This cooling period coincided with three major volcanic eruptions in 535/536, 539/540, and 547 CE (Büntgen et al. 2016). In Mediterranean regions, LAILA led to increased aridity and a notable reduction in forest cover, adversely affecting both evergreen and deciduous broadleaf trees (Corella et al. 2016; Michelangeli et al. 2022). It is plausible that these climatic changes impacted the abundance and distribution of falcon populations in Europe, as observed later during the Medieval Climate Anomaly (Hughes and Diaz 1994; Moseley et al. 2019). Unfortunately, there are very few bone remains from this period (see Appendix). Corbino and Albarella (2019) reported *Falco* sp. in an urban area of Italy from the fifth to sixth century CE, along with a common kestrel from a rural setting. Additionally, an unidentified *Falco* sp. was discovered in the Early Byzantine city of Caričin Grad, in present-day southern Serbia (Baron et al. 2019).

Before its abandonment in the mid-sixth century CE, the city of Berenike on the Red Sea Coast in Egypt was partially inhabited by the semi-nomadic Blemmyes, who originated from the Nubian region (Cobb 2021). Within the Falcon Shrine, approximately 65% of the faunal remains are falcons, most of which were beheaded for ritual purposes (see Fig. 3.4; Oller et al. 2022). These include sub-adult saker

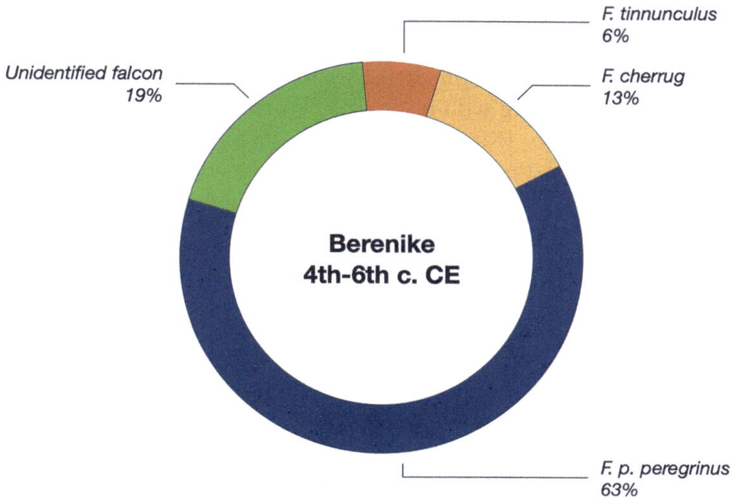

Fig. 3.4 Percentage distribution of falcons found in the Falcon Shrine at the Port of Berenike. (Red Sea, Egypt; data from Oller et al. 2022)

falcons and peregrines, with one adult falcon and a unique common kestrel, likely from captivity (Oller et al. 2022). Additionally, excavations revealed eggshells from falcons, providing clear evidence of direct access to their nests (Oller et al. 2022).

Among literary sources, Columella, a Greek philosopher and naturalist from the fourth to fifth century BCE, describes lesser kestrels predominantly nesting on the walls of buildings in Gades (modern-day Cádiz; Negro et al. 2020). Accordingly, a stylized depiction of a hovering kestrel appears on the base of a fifth century BCE Athenian drinking cup (Lamberton and Rotroff 1985). Columella uses the name "*titiunculus*," which Pliny translates as "*tinnungulus*," derived from "*tinnulus*" (shrill-sounding; Lindsay 1918; Jobling 2010). Pliny characterizes this bird as "*friendly*" to pigeons, as it protects them from other predators (Lindsay 1918). From an anthropomorphic approach, Pliny might have highlighted the advantages of pigeons nesting near lesser kestrel colonies (Campobello et al. 2015).

Umayyad interactions with Byzantium and Central Asia highlight a mutual fascination with high-flying and stooping peregrine and saker falcons, which were employed in the hunting of bustards, waterfowl, and *Columbiformes* (Strayer 1985).

Falcons are notably scarce at Early and Middle Anglo-Saxon sites in England (Crabtree 2016). The discovery of a female peregrine falcon skeleton at the Brandon Middle Anglo-Saxon site, combined with its deliberate burial, highlights the bird's importance and suggests it may have been a highly valued captive (Poole 2018). In contrast, smaller falcons like merlins and kestrels likely competed with humans for smaller prey (Yalden and Albarella 2009).

The cooling event (due to LAILA) led to reduced agricultural productivity, resulting in famines and migration from the Central Asian steppes (Büntgen et al. 2016). During the Migration period, the practice of hunting with falcons was introduced to Europe by the Goths (Epstein 1943; Birkhan 1972; Parker 1988). The floor mosaics in the House of the Falconer in Argos, Greece, provide the earliest known depictions of falconry, dating to the sixth century (Åkerström-Hougen 1974, 1981). References to hunting with birds of prey first appear in European writings from the mid-fifth century (Müller 1993; Stuhmiller 2015). The Germanic "*Lex Salica*," from the early sixth century, is the first legal code to regulate this practice (Stuhmiller 2015). "*Lex Baiuwariourum*" (sixth to eighth century CE) details raptors used in hunting: the "*cranohari*" for cranes, the "*canshpuh*" for wild geese, and the "*anothapuh*" for wild ducks (Müller 1993; Nótári 2013; see also Box 3.2).

Box 3.2: Falcon Names

Which modern species correspond to the "*cranohari*," "*canshpuh*," and "*anothapuh*" listed in the "*Lex Baiuwariourum*? (See Sect. 3.6). From this chapter onward, except for the final one, aligning historical names with species identified through modern taxonomy will continue to be a complex and often unresolved challenge. Ancient classifications of fauna frequently diverge from contemporary biological categories and may have also reflected regional variations. Additionally, the diverse raptors used in falconry highlight the ongoing issues with the nomenclature (Allsen 2006). Based on linguistic and phonetic analysis of the names in the *Lex Baiuwariourum*, it is likely that the lanner falcon is the species referred to as the duck hunter ("*anothapuh*"; Evans 1967). In the medieval falconry terminology, the lanner falcon is also known as "*blue legs*" (Dalby 1965). However, from an ecological or zoological standpoint, these terms might encompass multiple species with similar behaviors or physiological traits. For instance, the ancient Greek term ἱέραξ πελάγιος (sea-falcon) remains ambiguous. It could denote either Eleonora's falcon or the osprey (*Pandion haliaetus*), both of which are raptors commonly found in coastal regions. Aristotle's writings use "*Hierax*" (ἱέραξ) to refer to smaller diurnal raptors, excluding eagles and vultures, as long as they are about 60 cm or less in length (Arnott 2007). Modern derivatives of "*Hierax*." such as "*gerako*" and "*gerakina*," are used to denote falcons/hawks and buzzards, respectively. The broad usage of such terms, which cover various taxa, complicates the identification of specific plumage characteristics. For example, in the Byzantine falconry manual "*Orneosophion II*," it is challenging to believe that the term "*Hierakion*" (little *Hierax*) exclusively refers to the dark phase of Eleonora's falcon (Arnott 2007). Conversely, comparative studies can provide valuable insights by examining the etymology of these terms. François Viré (1922–1999), for instance, made significant strides in identifying the primary raptors mentioned in Arabic manuals, enhancing our understanding of Persian terminology (Allsen 2006). Thus, a multidisciplinary approach is essential for interpreting these terms from various perspectives and determining their biological significance (see also Sect. 3.1 and Box 3.3).

Box 3.3: Falcon Bone Assemblages and Their Relation to Human Use

During Pleistocene and Holocene, the presence of falcon species in assemblage correlated with their proximity to human populations (see Sect. 1.5). Neanderthals opportunistically captured small falcons while primarily focusing on more edible species (see Sect. 1.5). They likely used falcon skeletons, especially wings, symbolically by extracting feathers or raptor claws (see Sect. 1.5.1). Falcon hunting is challenging, except when capturing young or inexperienced birds (Kuhn et al. 2004). Some parts, such as feathers, claws, and talons, might have been scavenged from dead birds in the landscape rather than those actively hunted (Yeomans and Ritcher 2018). The opportunistic approach of humans to capturing or scavenging falcons provides valuable insights into their historical distribution and survival (see Sect. 2.4.2). Therefore, the presence or absence of falcons in fossil assemblages could directly reflect their encounter rates in the field (see Sect. 2.4.2).

In Ancient Egypt, falcon breeding became a commercial enterprise, supplying birds for the mummy industry (Warburton 2018; Ikram and Spitzer 2021). This practice significantly impacted the composition of bone assemblage, especially for common species like *F. tinnunculus*, which were easily reared in captivity (see Fig. 3.5). Falcons were likely acquired from nests, as it is easier to capture nestlings and fledglings than adults or to trap them during the migration period (Canby 2002).

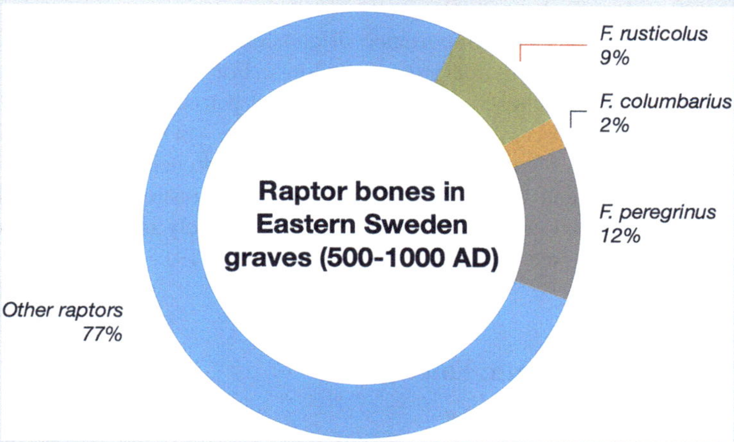

Fig. 3.5 Research conducted in Eastern Sweden has identified falconry birds in 34 aristocratic graves dating from CE 500 to 1000. Among the raptors discovered, the most frequent species is the goshawk (*Accipiter nisus*), which represents 73% of the total remains. (Data from Tyrberg 2002)

Falconry likely originated in Scandinavia and Central Europe around the mid-fifth to sixth century CE (Parker 1988; Tyrberg 2002; Allsen 2006; Smets and Abeele 2007; Gál 2015; Appendix). Graves from this period have yielded falconry-related remains, while earlier graves lack such artifacts (Tyrberg 2002; Fig. 3.4). These richly endowed graves reflect the aristocratic warrior ethos of the emerging Germanic states in the former Roman Empire (Tyrberg 2002). Consequently, warriors were buried with their horses and prized raptors, such as falcons (see Box 3.3 and Fig. 3.4; Tyrberg 2002; Gál 2015).

In Late Antiquity, the rise of falconry played a crucial role in the commercialization of falcons, a trend that intensified during the Medieval period. In Eastern Sweden, falcons, particularly gyr- and peregrine falcons, became symbols of aristocracy and were buried with their owners for the afterlife (see Sect. 3.6; Tyrberg 2002; Strehlau 2018). However, fossilized remains from graves dating between 500 and 1000 CE also include other raptors (Fig. 3.5). Analyzing the frequencies of these birds presents several interpretational challenges, including the possibility of exotic species being imported for falconry (Yalden and Albarella 2009). While the common kestrel is the most prevalent species in the region, it is notably absent from tombs (Fig. 3.5; Tyrberg 2002), which may suggest it was less valued in falconry. The goshawk is the most frequently found raptor in warrior burials, constituting over 70% of the raptor remains, likely due to the dense wooded areas of the Swedish Iron Age countryside (Tyrberg 2002). Some gyrfalcons and peregrine falcons have been found, but they are less common (Fig. 3.5). It is possible that gyrfalcons were either traded over long distances during the Iron Age or captured as passage migrants in southern Sweden (Tyrberg 2002).

This study emphasizes the importance of considering a species' intrinsic ecological characteristics when interpreting findings. For example, the gyrfalcon's potential for breeding in those areas and during that period was *a priori* excluded from consideration (Tyrberg 2002). To avoid inaccuracies in understanding the past distribution of falcons, a cautious, multidisciplinary approach is essential (Dirrigl et al. 2020; Leonardi and Sutton 2020). Additionally, reconstructing paleoclimatic and paleoenvironmental conditions is crucial for accurately interpreting the presence of fossil remains (see for instance, Moseley et al. 2019).

3.7 Western Palearctic Falcons

Reconstructing the past distributions of individual species is not feasible with the limited data available for this period (Appendix). Additionally, it is essential to differentiate between items sourced from trade for falconry and those from other origins. Nevertheless, it is possible to outline potential distributions by classifying falcons into two broad categories: small and large.

3.7.1 Small Falcons

The distribution range of small falcons demonstrates significant continuity with the latter part of the Early Holocene (see Chap. 2). Warm-climate species, such as the lesser kestrels and Eleonora's falcons, continued to be prevalent in Southern Europe and North Africa (Fig. 3.6). Merlins, which migrated to their current breeding regions following deglaciation, remain migratory at lower latitudes, including the Levant and Egypt (Fig. 3.6). Similarly, the Eurasian hobby's range extends from Sweden to Dagestan and Egypt (see Appendix; Fig. 3.6). As expected for a highly adaptable species like the common kestrel, it displays a broad distribution across the Western Palearctic (Fig. 3.6).

When examining the data sources for this reconstruction, it is essential to consider information from captive breeding, trade, and indirect sources like artifacts. For example, acknowledging the captive breeding of common kestrels in Ancient Egypt and the inferred presence of Eleonora's falcon in the Greek islands, as depicted in ancient frescoes, is crucial (Fig. 3.6). Additionally, the capture of these birds for various uses, such as funeral offerings or decoys, appears related to their abundance during breeding seasons and migrations. Similarly, in Spain, lesser kestrels utilize man-made structures for breeding (Fig. 3.6).

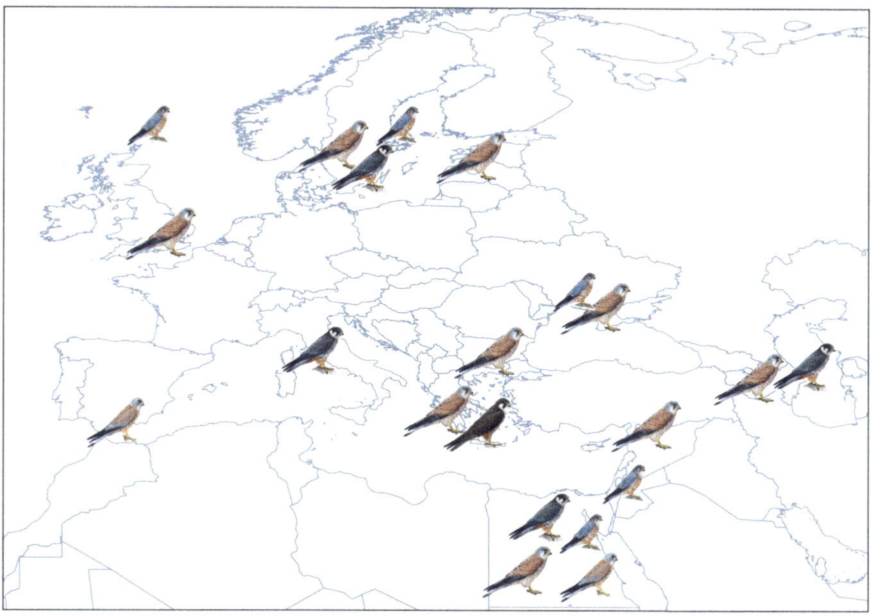

Fig. 3.6 The historical distribution of small falcons, including lesser and common kestrels, red-footed falcons, Eleonora's falcons, merlins, and Eurasian hobbies, has been reconstructed from archaeological remains and artifacts from the Ancient History and Late Antiquity periods. (Drawings courtesy of Marco Preziosi)

3.7.2 Large Falcons

The limited data available for large falcons during this historical period complicate efforts to fully reconstruct their distributions. However, several cautious observations can be made. The peregrine falcon appears to be widespread across the entire Western Palearctic, with captures reported at nesting sites and during migration (Fig. 3.7). In Egypt, it is important to consider both migratory individuals and the nesting presence of the subspecies *F. p. pelegrinoides* (barbary falcon). The two sightings of saker falcons are associated with captures during migration in Egypt. The lanner falcon, frequently observed in significant numbers (see Fig. 3.3), is linked to its use as a votive offering in pet cemeteries and temples in Ancient Egypt. This observation, excluding captive breeding, aligns with the species' presence in the same geographical area (Egypt and the Levant) during the Early Holocene (see Fig. 2.14). The gyrfalcon's presence in Southern Sweden does not correlate with the areas of discontinuous permafrost during the Pleistocene–Holocene transition (PHT; see Fig. 2.16). As suggested by Tyrberg (2002), the gyrfalcon may have been traded over long distances during the Iron Age or captured as passage migrants in southern Sweden (see Box 3.3).

For large falcons, a cautious and, above all, multidisciplinary approach is crucial for accurate data interpretation (see Sect. 3.1; Boxes 3.2 and 3.3). Unlike small falcons, these larger birds were actively used in the practice of falconry (see Chap. 4). They also hold significant cultural importance, symbolizing both temporal and religious power (see Box 3.1 and Chap. 4). Moreover, they are prominently featured

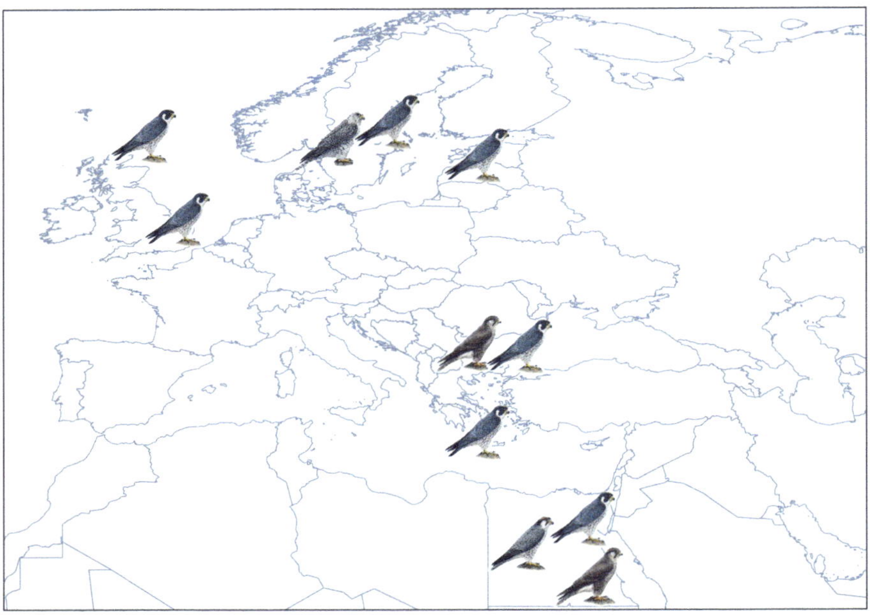

Fig. 3.7 The historical distribution of large falcons, including lanner, saker, and gyr- and peregrine falcons, has been reconstructed from archaeological remains and artifacts dating to the Ancient History and Late Antiquity periods. (Drawings courtesy of Marco Preziosi)

in poetry, art, and crafts and have a rich literary tradition related to falconry (see Chap. 4). Each of these sources provides valuable data that must be interpreted alongside zooarchaeological records and information on past climatic and environmental conditions (Fig. 3.7).

3.8 Conclusions

The Roman Warm Period (250 BCE to 400 CE) and the Late Antique Little Ice Age (sixth and seventh centuries CE) were two significant climate anomalies that shaped the first millennium of the Common Era (CE). It is plausible to hypothesize that these climatic changes influenced the distribution and density of falcons, primarily at a local scale. Similar impacts were observed during the ice movements of the third millennium BCE and throughout the Medieval Warm Period, affecting the nesting patterns of falcons such as gyrfalcons (Burham et al. 2009; Moseley et al. 2019).

Several key points can be highlighted:

- Captive breeding of small falcons (e.g., common kestrel), including practices like force-feeding, is a widely recognized procedure.
- Trade in falcons is documented, with large falcons primarily used for falconry and small falcons used as decoys.
- Evidence of nesting (e.g., lesser kestrels) on man-made structures has been observed.

Appendix: Fossil Remains (f) and Artifacts (a) of Western Palearctic Falcons from Ancient history to Late Antiquity (34th BCE to 6th CE)

Species	Dates	Location	References
F. naumanni[(f)]	1st millennium BCE (2nd–1st c.)	Dra Abu al-Naga, Egypt	Ikram and Spitzer (2021)
F. naumanni[(f)]	1st millennium BCE (2nd–1st c.)	Tinto Juan de la Cruz, Spain	Hernández (1993)
F. naumanni[(f)]	1st millennium BCE (2nd–1st c.)	Tuna e-Gebel, Egypt	Bailleul-LeSuer (2019)
F. naumanni[(f)]	1st millennium BCE (2nd–1st c.)	Kom Ombo, Egypt	Bailleul-LeSuer (2019)
F. tinnunculus[(f)]	4th millennium BCE (32nd c.)	Ajvide, Sweden	Mannermaa and Storå (2006)
F. tinnunculus[(f)]	1st millennium BCE (2nd–1st c.)	Dra Abu al-Naga, Egypt	Ikram and Spitzer (2021)
F. tinnunculus[(f)]	1st millennium BCE (2nd–1st c.)	Tuna e-Gebel, Egypt	Bailleul-LeSuer (2019)

(continued)

Species	Dates	Location	References
F. tinnunculus[(f)]	1st millennium BCE (2nd–1st c.)	Kom Ombo, Egypt	Bailleul-LeSuer (2019)
F. tinnunculus[(f)]	1st millennium BCE (2nd c.)	Portchester (Hants), UK	Parker (1988)
F. tinnunculus[(f)]	1st millennium BCE (2nd c.)	Barton Court Farm (Oxon), UK	Parker (1988)
F. tinnunculus[(f)]	1st millennium BCE (2nd c.)	Towcester (Northants), UK	Parker (1988)
F. tinnunculus[(f)]	1st millennium BCE (9–7th c.)	Khirbat al-Mudayna, Jordan	Popkin (2009)
F. tinnunculus[(f)]	1st millennium BCE (9–7th c.)	Tall Jawa, Jordan	Popkin (2009)
F. tinnunculus[(f)]	1st millennium BCE (10th c.)	Argishtihinili, Armenia	Manaseryan and Balyan (2002)
F. tinnunculus[(a)]	2nd millennium BCE (13th c.)	QV66, Valley of Queen, Egypt	Houlinan and Goodman (1986)
F. tinnunculus[(f)]	2nd millennium BCE (16th c.)	Yazılıkaya, Turkey	Boessneck (1975)
F. tinnunculus[(a)]	3rd millennium BCE	Akrotiri, Greece	Binnberg (2018)
F. tinnunculus[(f)]	3rd millennium BCE (22–20th c)	Tell Sweyhat, Syria	Buitenhuis (1985)
F. tinnunculus[(f)]	2nd millennium BCE (18–16th c)	Tell el-Dab'a, Egypt	Boessneck (1976)
F. tinnunculus[(f)]	3rd millennium BCE (33–28th c)	Velikent, Daghestan	Morales-Muñiz and Antipina (2003)
F. tinnunculus[(f)]	5th–6th c. CE	San Giovanni di Ruoti, Italy	Corbino and Albarella (2019)
F. tinnunculus[(f)]	2nd millennium BCE (17–13th c)	Kommos, Greece	Mylona (2022)
F. tinnunculus[(f)]	1–6th c. CE	Nicopolis-ad-Istrum, Bulgaria	Boev and Beech (2007)
F. tinnunculus[(f)]	1st millennium BCE	Howe, Orkney, UK	Yalden and Albarella (2009)
F. tinnunculus[(f)]	1st millennium BCE (4th–3th c)	Chajka, Ukraine	Gorobets and Rudenko (2022)
F. tinnunculus[(f)]	1st millennium BCE (10th–4th c)	Asva, Estonia	Maldre et al. (2018)
F. tinnunculus[(a)]	1st millennium BCE (5th c)	Athens, Greece	Lamberton and Rotroff (1985)
F. tinnunculus[(a)]	Early Bronze Age	Iwanowice, Poland	Bochenski et al. (2016)
F. vespertinus[(f)]	1st millennium BCE	Egypt	Fenech (2015)
F. vespertinus[(f)]	1st millennium BCE (2nd–1st c.)	Tuna e-Gebel, Egypt	Bailleul-LeSuer (2019)
F. eleonorae[(a)]	3rd millennium BCE	Akrotiri, Greece	Binnberg (2018)
F. columbarius[(f)]	1st millennium BCE (2nd–1st c)	Dra Abu al-Naga, Egypt	Ikram and Spitzer (2021)

(continued)

Species	Dates	Location	References
F. columbarius(f)	4–7th c. CE	Upper Zohar, Israel	Kroll (2012)
F. columbarius(f)	1st millennium BCE	Bu and Howe, Orkney, UK	Yalden and Albarella (2009)
F. columbarius(f)	Vendel Period (540–790 CE)	Valsgärde	Strehlau (2018)
F. columbarius(f)	1st millennium BCE (6th c)	Olbia Pontica, Ukraine	Gorobets and Rudenko (2022)
F. columbarius(a)	1st c. CE	Pompeii, Italy	Watson (2002)
F. subbuteo(f)	1st millennium BCE (1st c.)	Settefinestre, Italy	Rielly (1985)
F. subbuteo(f)	1st millennium BCE (2nd–1st c.)	Tuna e-Gebel, Egypt	Ikram and Spitzer (2021)
F. subbuteo(f)	1st millennium BCE (2nd–1st c.)	Kom Ombo, Egypt	Bailleul-LeSuer (2019)
F. subbuteo(f)	4th millennium BCE (32nd c.)	Ajvide, Sweden	Mannermaa and Storå (2006)
F. subbuteo(f)	3–4th c. CE	Alle, Les Aiges, Switzerland	Putelat (2010)
F. subbuteo(f)	1st millennium BCE (2nd–1st c)	Dra Abu al-Naga, Egypt	Ikram and Spitzer (2021)
F. subbuteo(f)	3rd millennium BCE (33–28th c)	Velikent, Dagestan	Morales-Muñiz and Antipina (2003)
F. biarmicus(f)	Hellenistic-Roman period	Elkab, Egypt	Gautier (2005)
F. biarmicus(f)	1st millennium BCE (2nd–1st c)	Dra Abu al-Naga, Egypt	Ikram and Spitzer (2021)
F. biarmicus(f)	1st millennium BCE (2nd–1st c)	Tuna e-Gebel, Egypt	Bailleul-LeSuer (2019)
F. biarmicus(f)	1st millennium BCE (2nd–1st c)	Kom Ombo, Egypt	Bailleul-LeSuer (2019)
F. cherrug(f)	Hellenistic-Roman period	Elkab, Egypt	Gautier (2005)
F. cherrug(f)	1st millennium BCE (2nd–1st c)	Tuna e-Gebel, Egypt	Bailleul-LeSuer (2019)
F. cherrug(f)	1st millennium BCE (2nd–1st c)	Kom Ombo, Egypt	Bailleul-LeSuer (2019)
F. cf. cherrug(f)	1st millennium BCE (12–4th c.)	Gala Abu Ahmed, Sudan	Linseele and Pöllath (2015)
F. cherrug(f)	6th c. CE	Kabyle, Bulgaria	Boev and Ribarov (1993)
F. rusticolus(f)	Late Antiquity (trade)	Central-Southern Sweden	Tyrberg (2002)
F. rusticolus(f)	ca. 1310–810 BCE	Nipisat I, Greenland	Gotfredsen and Møbjerg (2004)
F. peregrinus(f)	Hellenistic-Roman period	Elkab, Egypt	Gautier (2005)
F. peregrinus(f)	3rd millennium BCE (26th c.)	Tofts Ness, Scotland, UK	Serjeantson (2007)
F. peregrinus(f)	1st millennium BCE (2nd–1st c.)	Dra Abu al-Naga, Egypt	Ikram and Spitzer (2021)

(continued)

Species	Dates	Location	References
F. peregrinus[f]	1st millennium BCE (2nd–1st c.)	Tuna e-Gebel, Egypt	Bailleul-LeSuer (2019)
F. peregrinus[f]	1st millennium BCE (2nd–1st c.)	Kom Ombo, Egypt	Bailleul-LeSuer (2019)
F. peregrinus[f]	1st millennium BCE (4–1st c.)	Ust' Poluisk, Siberia	Potapova and Panteleyev 1999
F. peregrinus[f]	5th–6th c. CE	Brandon, UK	Poole (2018)
F. peregrinus[f]	2nd millennium BCE (17–13th c)	Kommos, Greece	Mylona (2022)
F. peregrinus[f]	1–6th c. CE	Nicopolis-ad-Istrum, Bulgaria	Boev (2006)
F. peregrinus[f]	1st millennium BCE	Howe, Orkney, UK	Yalden and Albarella (2009)
F. peregrinus[f]	1st millennium BCE (3rd–1st c.)	Ust' Poluisk, Siberia	Potapova and Panteleyev (1999)
F. peregrinus[f]	Vendel Period (540–790 CE)	Gnista, Sweden	Strehlau (2018)
F. peregrinus[f]	Vendel Period (540–790 CE)	Rickeby, Sweden	Strehlau (2018)
F. peregrinus[f]	1st c. CE	Berenice, Egypt	Osypinska et al. (2020)
F. peregrinus[f]	Migration Period (375–800 CE)	Salme, Estonia	Girininkas and Daugnora (2018)
Falco sp. [a]	1st millennium BCE (6th c.)	Amnisos, Crete	Heraklion Archeol Mus—Crete
Falco sp. [f]	2nd millennium BCE (14–11th c)	Saruq al-Hadid, UAE	Roberts et al. (2020)
Falco sp. [a]	4th millennium BCE (31st c.)	Hierakonpolis, Egypt	Egyptian Museum, Cairo, Egypt
Falco sp. [f]	3rd millennium BCE (22–20th c)	Phylakopi, Melos, Greece	Binnberg (2018)
Falco sp. [f]	6th c. CE	Mártély, Hungary	Gál (2015)
Falco sp. [f]	5th–6th c. CE	Carminiello ai Mannesi, Italy	Corbino and Albarella (2019)
Falco sp. [f]	6th c. CE	Caričin Grad, Serbia	Baron et al. (2019)
Falco sp. [f]	1st millennium BCE	Opatów, Kłobuck, Poland	Bochenski et al. (2016)

References

Åkerström-Hougen G (1974) The calendar and hunting mosaics of the Villa of the Falconer in Argos. Astrom, Stockholm

Åkerström-Hougen G (1981) Falconry as a motif in early Swedish art. Its historical and art historical significance. In: Zeitler R (ed) Les Pays du Nord et Byzance. Acta Universitatis Upsaliensis, Figura, Nova Series 19. Almqvist and Wiksell, Uppsala, pp 263–293

Allsen TT (2006) The royal hunt in Eurasian history. University of Pennsylvania Press, Philadelphia

Arnott WG (2007) Birds in the ancient world from A to Z. Routledge, Abingdon

Bailleul-LeSuer R (2019) From Egyptology to ornithology. The Cults of Sacred Falcons and The Musée des Confluences' Raptor Mummies. In: Porcier S, Ikram S, Pasquali S (eds) Creatures of Earth, Water and Sky. Essays on Animals in Ancient Egypt and Nubia. Sidestone Press, Leiden, pp 87–97

Baron H, Reuter AE, Marković N (2019) Rethinking ruralization in terms of resilience: subsistence strategies in sixth-century Caričin Grad in the light of plant and animal bone finds. Quat Int 499:112–128. https://doi.org/10.1016/j.quaint.2018.02.031

Barringer JM (2001) The hunt in Ancient Greece. Johns Hopkins University Press, Baltimore

Bibb R, McKnight L (2022) Identification of bird taxa species in ancient Egyptian mummies: Part 2, a qualitative evaluation of the utility of CT scanning and 3D printing. J Archaeol Sci Rep 46:103668. https://doi.org/10.1016/j.jasrep.2022.103668

Binnberg JK (2018) Birds in the Aegean Bronze Age. Doctoral dissertation, University of Oxford, Oxford

Birkhan H (1972) Wie kam die Falknerei nach Europa. Zeitschriftfür agdwissenschaft 18:152–166

Bochenski ZM, Tomek T, Wertz K, Wojenka M (2016) Indirect evidence of falconry in Medieval Poland as inferred from published zooarchaeological studies. Int J Osteoarchaeol 26:661–669. https://doi.org/10.1002/oa.2457

Boessneck J (1975) Die Tierknochen aus der Kammer C. In: Bittel K et al (eds) Yazılıkaya: Das hethitische Felsheiligtum Yazılıkaya. Boğazköy-Hattuša vol. 9, Berlin, p 62

Boessneck J (1976) Tell el-Dab'a III. Die Tierknochenfunde 1966–1969. In: Österreichische Akademie der Wissenschaften (ed) Denkschriften der Gesamtakademie, vol 5, Untersuchungen der Zweigstelle Kairo des Österreichischen Archäologischen Institutes. Verlag der Österreichischen Akademie der Wissenschaften, Kairo

Boev Z (2006) Late Holocene avian remains from the localities of the Roman period in Bulgaria. Hist Nat Bull 17:109–123

Boev Z, Beech MJ (2007) The bird bones. In: Poulter AG (ed) Nicopolis ad Istrum III: A late Roman and early Byzantine City: the finds and the biological remains. Oxbow, London, pp 242–253

Boev Z, Ribarov G (1993) Birds from the ancient town of Kabyle (1st Millennium B.C.—6th century A.D.). Historia Naturalis Bulgarica 4:68–77

Brehm AE (1858) Vorläufige Zusammenstellung der Vögel Spaniens mit kritischer Benutzung der bisher von spanischen Ornithologen herausgegebenen Verzeichnisse. Allgemeine Deutsche Naturhistorische Zeitung. Im Auftrage der Gesellschaft ISIS in Dresden NF 3:431–448; 449–489

Bryce T (2012) The Hittite Empire. In: Potts DT (ed) A companion to the archaeology of the ancient near East. Wiley-Blackwell, Oxford, pp 721–739

Buitenhuis H (1985) Animal remains from Tell Sweyhat, Syria. Palaeohistoria 25:131–144

Büntgen U, Myglan VS, Ljungqvist FC, McCormick M, Di Cosmo N, Sigl M, Jungclaus J, Wagner S, Krusic PJ, Esper J, Kaplan JO (2016) Cooling and societal change during the Late Antique Little Ice Age from 536 to around 660 AD. Nat Geosci 9(3):231–236. https://doi.org/10.1038/ngeo2652

Burham KK, Burham WA, Newton I (2009) Gyrfalcon *Falco rusticolus* post-glacial colonization and extreme long-term use of nest-sites in Greenland. Ibis 151:514–522. https://doi.org/10.1111/j.1474-919X.2009.00939.x

Cameron A (2012) The Mediterranean World in Late Antiquity 395-700 AD. Routledge, Oxon

Campagno M (2013) Late fourth millennium BCE. In: Grajetzki W, Wendrich W (eds) UCLA encyclopedia of egyptology, Los Angeles

Campobello D, Hare JF, Sarà M (2015) Under my wing: lesser kestrels and jackdaws derive reciprocal benefits in mixed-species colonies. Behav Ecol 23:425–433. https://doi.org/10.1093/beheco/arr207

Canby JV (2002) Falconry (Hawking) in Hittite Lands. J Near East Stud 61:161–201

Cobb M (2021) Barbarians and Blemmyes: who was in control of the Red Sea Port of Berenike in the late Antique period. J Late Antiquity 14:267–293. https://doi.org/10.1353/jla.2021.0029

Corbino CA, Albarella U (2019) Wild birds of the Italian Middle Ages: diet, environment and society. Environ Archaeol 24(4):449–459. https://doi.org/10.1080/14614103.2018.1516371

Corella JP, Valero-Garcés BL, Vicente-Serrano SM, Brauer A, Benito G (2016) Three millennia of heavy rainfalls in Western Mediterranean: frequency, seasonality and atmospheric drivers. Sci Rep 6:38206. https://doi.org/10.1038/srep38206

Crabtree PJ (2016) Zooarchaeology at medieval Ipswich: from wic to regional market town. In: Jervis B, Broderick LG, Sologestoa IG (eds) Objects, environment, and everyday life in Medieval Europe. Brepols, Turnhout, pp 19–40

Dalby D (1965) Lexicon of the medieval German hunt. De Gruyter, Berlin

Darnell JC (2002) Theban Desert Road Survey in the Egyptian Western Desert, vol 1: Gebel Tjauti rock inscriptions 145 and Wadi el-Hôl rock inscriptions 1–45. Oriental Institute Publication 119, The University of Chicago Press, Chicago

Dirrigl FJ, Brush T, Morales-Muñiz A et al (2020) Prehistoric and historical insights in avian zooarchaeology, taphonomy and ancient bird use. Archaeol Anthropol Sci 12:57. https://doi.org/10.1007/s12520-020-01016-2

Edgerton W, Wilson J (1936) Historical records of Ramses III: the texts in Medinet Habu Volumes I and II. SAOC 12. The Oriental Institute of the University of Chicago, Chicago

Epstein HJ (1943) The origin and early history of falconry. Isis 34:497–509

Ericson PGE, Tyrberg T (2004) The early history of the Swedish avifauna. KVHAA, Stockholm

Evans D (1967) Lanier, historie d'un mot. Librairie Droz, Geneve

Fenech N (2015) Raptors and Falconry in Malta: Culture, history and conservation. In: Szymak U, Sianko P (eds) Falconry – its influence on biodiversity and cultural heritage. Muzeum Podlaskie, Białystok, pp 121–168

Foxhall L (1995) Bronze to iron: agricultural systems and political structures in late Bronze Age and early Iron Age Greece. Annu Br Sch Athens 90:239–250

Gál E (2015) "Fine feathers make fine birds": the exploitation of wild birds in Medieval Hungary. Antaeus 33:345–368

Gautier A (2005) Animal mummies and remains from the necropolis of Elkab (Upper Egypt). Archaeofauna 14:139–170

George A (2007) Babylonian and Assyrian: a history of Akkadian. In: Postgate JN (ed) Languages of Iraq, ancient and modern. Cambridge University Press, Cambridge, pp 31–71

Girininkas A, Daugnora L (2018) Premodern hunting with birds of prey in the historical Lithuanian lands: entertainment, politics or economy necessity? In: Gersmann K-H, Grimm O (eds) Raptor and human—falconry and bird symbolism throughout the millennia on a global scale—vol 1/3. Wachholtz, Kiel, pp 1215–1227

Gorobets L, Rudenko N (2022) Remains of birds from the Northwest Black Sea ancient settlements. Environ Archaeol 27:306–322. https://doi.org/10.1080/14614103.2021.1905933

Gotfredsen AB, Møbjerg T (2004) Nipisat—a Saqqaq culture site in Sisimiut, central west Greenland. Meddelelser om Grønland Man & Society 31. Danish Polar Center, Copenhagen

Gualtieri TFL (2005) Birds of prey and the sport of falconry in Italian literature through the Fourteenth Century: from serving love to served for dinner. PhD thesis, University of Wisconsin, Madison

Hernández F (1993) Catálogo provisional de los yacimientos con aves del cuaternario de la Península Ibérica. Archaeofauna 2:231–275

Houlihan PF, Goodman SM (1986) The birds of the ancient Egypt. A Phillips, Warminister

Huges MK, Diaz HF (1994) Was there a "Medieval Warm Period", and if so, where and when? Clim Chang 26:109–142. https://doi.org/10.1007/BF01092410

Ikram S, Spitzer M (2021) The cult of Horus and Thoth: a study of Egyptian animal cults in Theban tombs 11, 12, and -399. In: Daujat J, Hadjikoumis A, Berthon R, Chahoud J, Kassianidou V, Vigne J-D (eds) Archaeozoology of Southwest Asia and adjacent areas XIII. Proceedings of the thirteenth international symposium. Lockwood Press, Atlanta, pp 245–267

Ikram S, Slabbert R, Cornelius I, du Plessis A, Swanepoel LC, Weber H (2015) Fatal force-feeding or gluttonous gagging? The death of Kestrel SACHM 2575. J Archaeol Sci 63:72–77. https://doi.org/10.1016/j.jas.2015.08.015

Jobling JA (2010) The Helm dictionary of scientific bird names. C. Helm, London

Kenning J (2002) Zum Begriff s3b-šwt—ein Zugang aus der Falknerei. Zeitschrift für ägyptische Sprache und Altertumskunde 129:43–48

Kroll H (2012) Animals in the Byzantine Empire: an overview of the archaeozoological evidence. Archeol Medieval 39:93–121

Kuhn SL, Belfer-Cohen A, Barzilai O, Stiner MC, Kerry KW, Munro ND, Bar- Yosef Mayer D (2004) The last glacial maximum at Meged rockshelter, Upper Galilee, Israel. J Israel Prehist Soc 34:5–47

Lamberton RD, Rotroff SI (1985) Birds of the Athenian agora. American School of Classical Studies at Athens, Princeton

Leonardi G, Sutton LJ (2020) *Falco lanarius*: the contorted history of a poetical archetype and a taxonomic puzzle. Archaeol Anthropol Sci 12:22. https://doi.org/10.1007/s12520-019-00995-1

Lindsey WM (1918) Bird-names in Latin glossaries. Class Philol 13:1–22

Linglin M, Amiot R, Richardin P, Porcier S, Antheaume I, Berthet D, Grossi V, Fourel F, Flandrois J-P, Louchart A, Martin JE, Lécuyer C (2020) Isotopic systematics point to wild origin of mummified birds in Ancient Egypt. Sci Rep 10:15463. https://doi.org/10.1038/s41598-020-72326-7

Linseele V, Pöllath N (2015) Local foods and traded goods—the faunal remains from a Napatan outpost at Gala Abu Ahmed (Sudan, First Millennium BC). Afr Archaeol Rev 32:537–590. https://doi.org/10.1007/s10437-015-9194-7

Lucy S (2001) Roman Iron Age. In: Peregrine PN, Ember M (eds) Encyclopaedia of prehistory, vol 4. Springer, New York, pp 264–278

Maldre L, Tomek T, Peets J (2018) Birds of prey from Vendel Age ship burials of Salme (c. 750 AD) and in Estonian archaeological material. In: Gersmann K-H, Grimm O (eds) Raptor and human—falconry and bird symbolism throughout the millennia on a global scale—vol 1/3. Kiel, Wachholtz, pp 1229–1250

Manaseryan N, Balyan L (2002) The birds of ancient Armenia. Acta Zool Cracov 45:405–414

Mannermaa K, Storå J (2006) Stone Age exploitation of birds on the Island of Gotland, Baltic Sea. Int J Osteoarchaeol 16:429–452. https://doi.org/10.1002/oa.884

Manning JG (2018) The Open Sea: the economic life of the ancient Mediterranean world from the Iron age to the rise of Rome. Princeton University Press, Princeton

Michelangeli F, Di Rita F, Celant A, Tisnérat-Laborde N, Lirer F, Magri D (2022) Three millennia of vegetation, land-use, and climate change in SE Sicily. Forests 13:102. https://doi.org/10.3390/f13010102

Morales-Muñiz A, Antipina Y (2003) A preliminary survey of the birds from Velikent (Bronze Age, Daghestan, Russia). Bull Fla Mus Nat Hist 44:43–54

Morgan LW, McGovern-Huffman S (2008) Noninvasive radiographic analysis of an Egyptian falcon mummy from the late period 664-332 BC. J Avian Biol 39:584-587. https://doi.org/10.1111/j.2008.0908-8857.04269.x

Moseley GE, Rosvold J, Gotfredsen AB, Hajdas I, Gilg O, Gregersen KM, Spötl C, Edwards RL (2019) First pre-modern record of the Gyrfalcon (*Falco rusticolus*) in Northeast Greenland. Polar Res 38:3539. https://doi.org/10.33265/polar.v38.3539

Müller H-H (1993) Falconry in central Europe in the Middle Ages. In: Exploitation des animaux sauvages a travers le temps. Editions APDCA, Juan-les-Pines, pp 431–437

Murname W (1995) The history of ancient Egypt. In: Sasson J (ed) Civilizations of the ancient near East. Scribner's, New York, pp 691–717

Mylona D (2022) The Bronze Age birds in Greece. A zooarchaeological perspective. Quat Int 626–627:71–79. https://doi.org/10.1016/j.quaint.2020.11.001

Negro JJ, Prenda J, Ferrero JJ, Rodríguez A, Reig-Ferrer A (2020) A timeline for the urbanization of wild birds: the case of the lesser kestrel. Quat Sci Rev 249:106638. https://doi.org/10.1016/j.quascirev.2020.106638

Nótári T (2013) Bavarian linguistic elements in *Lex Baiuvariorum*. Acta Juridica Hungarica 54:272–285

Oller J, Fernández D, Trevín V, Kaper OE, Ast R, Osypińska M, Sidebotham SE (2022) A Falcon Shrine at the Port of Berenike (Red Sea Coast, Egypt). Am J Archaeol 126:567–591. https://doi.org/10.1086/720806

Osypinska M, Skibniewski M, Osypinski P (2020) Ancient pets. The health, diet and diversity of cats, dogs and monkeys from the Red Sea port of Berenice (Egypt) in the 1st-2nd centuries AD. World Archaeol 52:639–653. https://doi.org/10.1080/00438243.2020.1870545

Parker A (1988) The birds of Roman Britain. Oxf J Archaeol 7:197–226

Podgórski T (2010) Some remarks on the problem of the Horus falcon identification. Stud Ancient Art Civilization 14:67–72

Poole K (2018) Zooarchaeological evidence for falconry in England, up to AD 1500. In: Gersmann K-H, Grimm O (eds) Raptor and human—falconry and bird symbolism throughout the millennia on a global scale—vol 1/3. Wachholtz, Kiel, pp 1027–1053

Popkin RW (2009) The society and economy of Iron Age Transjordan: a contextual zooarchaeological analysis. PhD thesis, University College London, London

Porter R (2011) Insights into Egyptian horus falcon imagery by way of real falcons and horus falcon influence in the aegean in the middle bronze age: Part I. J Ancient Egyptian Interconnect 3(3):27–38

Potapova OR, Panteleyev AV (1999) Birds in the economy and culture of early Iron Age inhabitants of Ust' Poluisk, Lower Ob' river, Northwestern Siberia. Smithson Contrib Paleobiol 89:129–137

Putelat O (2010) Archéozoologie. In: Demarez J-D, Othenin-Girard B (eds) Établissements ruraux de La Tène et de l'Époque romaine à Alle et à Porrentruy. Cahiers d'Archéologie Jurassienne 28, pp 321–348

Rielly K (1985) Gli Uccelli. In: Ricci A (ed) Settefinestre, Una villa schiavistica nell'Etruria Romana, vol III. Panini, Modena, pp 299–302

Roberts J, Weeks L, Fillios M, Cable C, Fillios M, al-Aali YY et al (2020) The bird remains from Saruq al-Hadid: Insights into human activity and the environment in late prehistoric southeastern Arabia. Quat Int 543:71–76. https://doi.org/10.1016/j.quaint.2020.03.021

Serjeantson D (2007) Bird bones. In: Dockrill SJ, Bond JM, Smith AN, Nicholson RA (eds) Investigations in Sanday Orkney Volume 2: Tofts Ness Sanday, An Island Landscape Through 3000 Years of Prehistory. The Orcadian in association with Historic Scotland, Kirkwall, Orkney, pp 216–277

Sherratt S, Sherratt A (1993) The growth of the Mediterranean economy in the early first millennium BC. World Archaeol 24:361–378

Shi F, Sun C, Guion A, Yin Q, Zhao S, Liu T, Guo Z (2022) Roman warm period and late Antique little Ice Age in an earth system model large ensemble. J Geophys Res Atmos 127:e2021JD035832. https://doi.org/10.1029/2021JD035832

Smets A, Abeele BVD (2007) Medieval hunting. In: Resl B (ed) A cultural history of animals in the Medieval Age. Berg, Oxford/New York, pp 59–79

Strayer JR (1985) Dictionary of the Middle Age, vol 6. C Scribner's Sons, New York

Strehlau H (2018) Animals in burial contexts. MSc thesis. Uppsala University, Uppsala

Stuhmiller J (2015) Hunting, hawking, fowling, and fishing. In: Classen A (ed) Handbook of Medieval Culture, vol 2. De Gruyter, Berlin, pp 697–721

Tyrberg T (2002) The archaeological record of domesticated and tamed birds in Sweden. Acta Zool Cracov 45:215–231

von den Driesch A, Kessler D, Steinmann F, Bertelaux V, Peters J (2006) Mummified, deified and buried at Hermopolis Magna—the sacred birds from Tuna El-Gebel, Middle Egypt. Ägypten und die Levante 15:203–244

Wang T, Surge D, Mithen S (2012) Seasonal temperature variability of the Neoglacial (3300–2500BP) and Roman Warm Period (2500–1600BP) reconstructed from oxygen isotope ratios of limpet shells (*Patella vulgata*), Northwest Scotland. Palaeogeogr Palaeoclimatol Palaeoecol 317–318:104–113. https://doi.org/10.1016/j.palaeo.2011.12.016

Warburton DA (2018) Egypt and earlier: birds of prey in the human mind at the dawn of history. In: Gersmann K-H, Grimm O (eds) Raptor and human. Wachholtz, Hamburg, pp 325–346

Watson GE (2002) Birds: evidences from wall paintings, mosaics, sculpture, skeletal remains, and ancient authors. In: Jashemski WF, Meyer FG (eds) The natural history of Pompeii. Cambridge University Press, Cambridge, pp 357–400

Wilkinson JG (1878) The manners and customs of the ancient Egyptians. Scribner and Welford, New York

Wilson R (1983) Piazza Armerina. Granada, London

Yalden DW, Albarella U (2009) The history of British birds. Oxford University Press, Oxford

Yeomans L, Ritcher T (2018) Exploitation of a seasonal resource: bird hunting during the Late Natufian at Shubayqa 1. Int J Osteoarchaeol 28:95–108. https://doi.org/10.1002/oa.2533

Chapter 4
Middle Ages

4.1 Introduction

The Middle Ages provide valuable information about falcons in Europe, emphasizing their role in culture, art, poetry, and the history of falconry. However, it is important to assess how useful this information is for understanding the past distribution and population of falcons. The main sources of information on falcons are: (1) fossils, (2) bestiaries, (3) falconry treatises, and (4) management catalogues. Generally, these sources are not easy to compare and can offer very different information. For example, bestiaries do not provide detailed information about these predators, and none of them includes a chapter on *Falco* species (George and Yapp 1991). The difference in naming is clear in these relatively contemporary sources. Bestiaries referring to "*capus*" (falcon) drew from Isidore of Seville (7th c. CE) due to its curved toes, but it was indistinguishable from other raptors (George and Yapp 1991; Zambon 2018; Goldberg 2020). They also included "*aliaetus*" (possibly merlin, sparrowhawk, or small falcon in general) and "*herodius*" (a type of falcon, primarily the gyrfalcon, and also some *Accipiter* species; George and Yapp 1991). In contrast, during the thirteenth century, Emperor Frederick II of Hohenstaufen (1194-1250) used the term *Falco* for peregrine falcons and related birds (George and Yapp 1991). Considering the ambiguity and confusion surrounding these terms (as well as the illustrations in medieval manuscripts), Frederick II's work serves as a crucial reference point (see Sect. 4.3.1; Yapp 1979). His Latin work "*De Arte Venandi Cum Avibus*," was widely cited in later falconry treatises and bestiaries (see Sect. 4.2.1; Oggins 1993). Falcons and hawks likely began appearing in bestiaries around the time of the Crusades, owing to increased contact with the Arab world (Oggins 1993). However, Frederick II's empirical and scientific perspective stands out as an exception. During this historical period, the poetic, social, and cultural archetype broadly identified these iconic species. In the "*Booke of Hawkyng after Prince Edwarde Kyng of Englande*," later adapted in the compilation "*Boke of Seynt*

G. Leonardi, *The History of Western Palearctic Falcons*,
https://doi.org/10.1007/978-3-031-86528-2_4

Albans" (1486), various raptor species were placed within the medieval social hierarchy (Cummins 1988; Pickering and O'Mara 1999; Coy 2009). Yapp (1982), however, highlights inconsistencies in the texts, suggesting that the allocation of raptor species to medieval social ranks lacks a basis in historical reality.

Alongside the literary world, there existed a practical realm focused on profitable trades, diplomatic exchanges, and the management of falcons. For instance, gyrfalcons, mainly brought from northern countries to central Europe, were renowned for the specimens from Iceland (see Sect. 4.2.3; Müller 1993). Fossils in medieval trading hubs around the Baltic Sea, such as Reric, Hedeby, and Birka, show that northern European merchants sold raptors extensively (Goldberg 2020). This also indicates that during medieval times, birds of prey were often traded and transported over long distances. Additionally, raptors may have been captured for their feathers to fletch arrows (Müller 1993). Trading and keeping falcons required rules, but these rules often lacked specificity in identifying even broad systematic groups. During the Merovingian period, authors usually referred to all raptors as "*accipitres*." However, in the Carolingian era, a clearer distinction between hawks and falcons emerged. For example, early Carolingian reform councils in the 740s banned churchmen from hunting with both hawks and falcons (Goldberg 2020). Wild falcons had to be captured in their natural habitat and subsequently cared for and trained by specialized personnel who are compensated for their expertise (Fiorini 2001). Attendants may have selected *juvenile* falcons from nests and trained them for falconry (Serjeantson 2009). Frederick II provides a detailed account of the various stages at which falcons can be captured: as nestlings, as fledged birds, during postbreeding movements, in winter, and during prebreeding movements (Trombetti Budriesi 2000). In Charlemagne's capitularies, hawks (likely goshawks), falcons (probably peregrine falcons), and sparrowhawks were listed as the primary hunting birds (Goldberg 2020). In the collection of letters and administrative documents from Frederick II's reign, various mentions discuss the management of his falcons in Sicily and Apulia (Huillard-Bréholles 1852–1861). These documents also confirm that he sent his falconers, like Stefano Mustaccius, to collect birds, mainly in Sicily, the nearby islands, and in Malta (Bresc 1980). Lords could establish special restricted areas to protect the breeding grounds of birds of prey used in falconry (Arabadhiev 1962). Alongside falcon management, there was also a concern for prey populations. For example, King John (1199–1216) banned falconry in England for a season to allow game stocks to recover (Evans 1992).

Examining the fossils provides a more cohesive representation of the reliable densities and distributions of falcons in Europe (Fig. 4.1). Bones of predatory birds have been found in various medieval settlements across Europe, including castles and royal palaces (Appendix; Müller 1993; Mulkeen and O'Connor 1997). Presumably, the frequencies of these bones can be influenced by the abundance of prey, the use of urban debris, and the proximity to preferred natural habitats (Fig. 4.1; Mulkeen and O'Connor 1997). However, comparing different geographical regions reveals a consistent occurrence percentage that is not biased by a high number of individuals due to falconry practice (see also Box 3.3). Therefore, it is crucial to determine whether certain species were introduced to an area or if hypotheses about

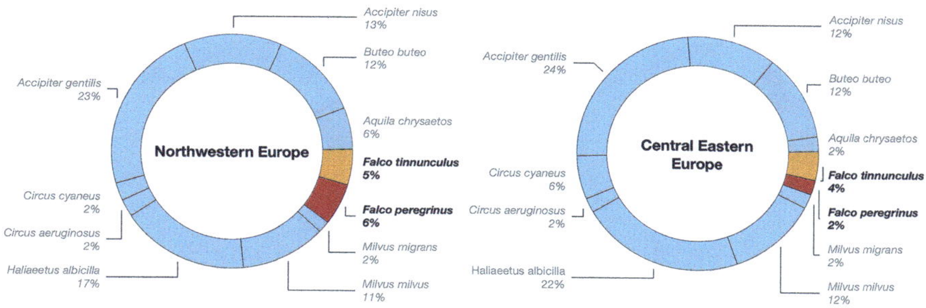

Fig. 4.1 Frequencies of raptors in bone material analyzed from medieval settlements in Northwestern Europe (28 sites) and Central Eastern Europe (32 sites; data from: Müller 1993; Mulkeen and O'Connor 1997)

past distributions lack support from a multidisciplinary approach (cf. Sects. 4.4.1 and 4.4.2).

This chapter covers the entire Middle Ages, from the seventh century to the fifteenth century CE (approximately 900 years). Despite numerous debates and alternative periodization systems, the traditional Middle Ages framework remains prevalent (Landro 2019). Romance language historians typically divide the Middle Ages into two parts: *High* (7–10th c. CE) and *Low* (11–15th c. CE). In contrast, English-speaking historians, influenced by German scholarship, use a three-part division: *Early* (7–10th c. CE), *High* (11–13th c. CE), and *Late* (14–15th c. CE). As in the preceding chapter, the periodization of the Middle Ages is shaped by human socioeconomic events, with minimal impact on falcon population ecology. Therefore, a straightforward division into *High* and *Low* periods will be adopted, supplemented with detailed insights in various sections.

4.2 High Middle Age (7th–10th c. CE)

During the High Middle Ages, large falcons like the peregrine falcon became crucial as status symbols. In Middle Saxon times, the presence of falcon bone remains suggests that the studied site was likely occupied by a relatively prosperous population (Crabtree 1996). Falcons were traded and also used as diplomatic gifts. The letters of Saint Boniface, a Devonshire native involved in the conversion of the Germans, reveal his role as a provider of "*accipitres*" and "*falcones*" to Aethelbald of Mercia and Aethelbert of Kent (Yapp 1982). Between 745 and 746, the archbishop gave two falcons to King Ethelbald of Mercia, and, from 748 to 755, King Ethelbert II of Kent asked Boniface for two falcons (possibly gyrfalcons; Dobney and Jaques 2002). "*Duos falcones quorum ars et artis audatia sit grues velle libenter captando arripere et arripiendo consternere solo*" translates to "*A pair of falcons of such cleverness and courage that they will without hesitation attack cranes, and, having caught them, will bring them to earth*" (Tangl 1916; Emerton and Noble

Fig. 4.2 Frequencies of raptors in bone material analyzed from funeral contexts in Sweden (9–12th c. CE) and urban settlements in Poland (mainly 9–10th) and Denmark (7–11th c. CE; Sten and Vretemark 1988; Gotfredsen 2014; Bochenski et al. 2018)

2000). These falcons were important for the aristocracy in the High Middle Ages, as evidenced by their inclusion in burial items. Between the seventh and tenth centuries, at least four peregrine falcons were discovered in chieftains' graves in Sweden (Sten and Vretemark 1988; Vretemark 2018; Fig. 4.2). The bones show that animals were sacrificed and buried with the deceased as part of their burial items. A notable assemblage includes two peregrines, one goshawk, and one sparrowhawk found at Rickeby (Uppland), dating back to the early seventh century (Vretemark 2018). Gyrfalcons are rare in falconry graves, likely because they were traded from northern regions (see Appendix; Vretemark 2018). A merlin was discovered in a Swedish grave, but common kestrels, regularly found in Danish settlements, have not been found in any graves in Sweden (Fig. 4.2).

4.2.1 Medieval Warm Period (MWP)

The Medieval Warm Period (MWP) was a warm climate phase in the North Atlantic region from around 950 to 1250 (Anchukaitis and Smerdon 2022). However, the concept of the MWP is overly simplistic, as historical data reveal the North Atlantic climate during this period was highly variable (Hartman et al. 2017). Additionally, temperatures in medieval Europe were comparable to, or possibly warmer than, those in the twentieth century (Luterbacher et al. 2016). In very northern areas like

Greenland, the mild climate has created favorable conditions for both wildlife and human occupation (Hartman et al. 2017; Borreggine et al. 2023). The discovery of fossilized gyrfalcon remains in a cave in Northeast Greenland indicates that the warmer temperatures during the MWP prevented substantial ice formation in the area (Appendix; Moseley et al. 2019).

4.2.2 Falconry Treatises

There is a remarkably broad body of academic literature on medieval falconry treatises, encompassing history, language, economics, politics, and prominent figures of the era. However, the focus on the birds' biology and environment is limited and sometimes inconsistent. Behavioral and ecological considerations are primarily discussed in relation to the birds' performances during practical field training. Ultimately, these works serve as manuals for feeding, training, caring for, and managing captive animals. The main exception to these manuals is Frederick II's "*De arte venandi cum avibus*" (see Sect. 4.3.1). Nevertheless, these sources, along with fossil evidence, offer several subtle clues to help understand the status of falcons in the Middle Ages.

During the Middle Ages, the scientific languages were Greek and Latin. Vernacular languages had to use elaborate lexical fields to express scientific terms (Lambert 2008). As many texts about falcons are in Arabic, translation becomes necessary (see Sect. 4.3.1). On the other hand, many terms used by Arab falconers, especially from the Maghreb, have Romance origins (Viré 1965). For instance, "*tarshūn*" in Arabic, derived from the distortion of "*tarshūl/ṭarshūl*," corresponds to "*torzuelo/terçuelo*," meaning "*male hunting bird*" (Viré 1965). In the "*Kitāb Ḍawari aṭ-ṭayr*" (8th c.) by al-Ġitrīf ibn Qudāmah al-Ghassānī and in the "*Kitāb al-kāfī fī al-bayzarah*' (10th c.), the use of a Romance loanword to designate the barbary falcon is "*balanjrīniyya*," which is the Arabic for the Latin "*pelegrinus*" (Viré and Möller 2002). When al-Ghitrīf describes the different species of falcons according to their color, he states that the black peregrines are the "*baḥriyya*." These are those "*qui naissent et croissent dans les îles bordant les littoraux*" (Viré 1966; Viré and Möller 2002). In the 1240 treatise "*al-Djamhara*" by 'Isa al-Asadī from Baghdad, there is a description of how children were skilled at training small falcon-like birds, such as kestrels, to hunt larks and sparrows (Trombetti Budriersi 2000). Another popular game of that time was flying merlins to chase small field birds (Trombetti Budriesi 2000).

One of the most important and renowned Arabic treatises is the "*Kitāb al-Mutawakkilī*," dedicated to the Abbasid Caliph al-Mutawakkil, authored by Abū Zayd Ḥunayn ibn Isḥāq al-Ibādī in the mid-ninth c. (Akasoy and George 2005). This text formed the basis for a significant portion of the Latin compilation known as "*Moamin*," a corruption of the phrase "*amīr al-mu'minin*," meaning "*prince of the believers*" (Akasoy and Georges 2005; Martinez Enamorado 2011). In the Franco-Italian version by Daniele Deloc da Cremona (1249–1272; permanent link: https://

www.rialfri.eu/texts/moamin|001), Arabic is used in highly specialized contexts, such as the nomenclature of falcons (Marruncheddu 2008). Five falcons are identified, along with three morphs of the saker falcon. Theodore of Antioch, in his Latin rendition of "*Moanim*," appears to adapt the Arabic names "*anūqī*" and "*anīqī*" as "*aynoki*" and "*aynoli*," which were later translated into French as "*aynolu*" (Marruncheddu 2008). Marruncheddu (2008) suggests that this may refer to the lanner falcon, but in one Italian version, it is identified as the "*smiriglio*" (merlin).

Latin treatises, such as those by *Guillelmus falconarius* and *Dancus Rex*, categorized falcons based on their dispositions: black feathers were linked to a melancholy temperament, white feathers with a phlegmatic disposition, and red with a sanguine nature (Oggins 1993). In contrast, ancient Arab writers classified hunting birds into two groups: those with yellow eyes, including hawks, harriers, and eagles, and those with black eyes, such as sakers, peregrines, lanners, and merlins (Trombetti Budriesi 2000). Nomadic Arabs trained the peregrine falcon, lanner falcon, saker, and merlin, all of which are birds of prey with black eyes. The large northern gyrfalcon, highly esteemed by Frederick II, was known primarily through reputation (Trombetti Budriesi 2000). Indeed, al-Ġitrīf does not mention the gyrfalcon in his hunting treatise, and it is not referenced in the ninth-century falconry book of the caliph al-Mutawakkil (Buquet 2021). Similarly, the falconer of the caliph al-ᶜAzīz, also omits the gyrfalcon in his tenth-century falconry book (Buquet 2021).

In addition to falconry manuals, there is a growing interest in the habitats, behaviors (especially hunting habits), and classifications of falcons. Written around 1188 by Giraldus Cambrensis (Gerald of Wales), "*Topographia Hibernica*" provides an account of the landscape and people of Ireland. Cambrensis noted on several occasions that gyr- and lanner falcons were not found in Ireland. Conversely, he considered merlins (*meruli minuti*) and hobbies (*hobeli aestivales*) to be the most common falcon species in the country. Brunetto Latini (c. 1220–1294), an Italian philosopher and scholar, categorized falcons in his work "*Il Tesoretto*" (The Little Treasure), written in Italian verse. He divided them into different families: "*Faucon lanier*," "*pelerins*," "*faucons monta[n]is*," "*faucons gentils ou gruier*," and "*le gerfal*" (Zambon 2018). Identifying other falcon species can be challenging, especially when distinguishing small falcons collectively known as "*merlins*," which come in grey or black rump variations (Zambon 2018). Despite these challenges, the practice of categorizing falcons dates back to before the twelfth century. The "*Epistola Aquilae Symachi et Theodotionis ad Ptolomeum regem Aegypti de avibus nobilibus*" was a widely disseminated text throughout medieval Europe (Fig. 4.3). Probably written in Northwestern Europe, it was summarized and reorganized in the naturalistic encyclopedia "*Liber de natura rerum*" by Thomas of Cantimpré around 1240 (Fradejas Rueda 2009). The section "*De diversis generibus falconum*" distinguished eight types of falcons: the genus "*lanarium*," "*peregrinum*," "*pedes ceruleum*," "*montanus*," the "*herodius*" (also known as "*gyrofalco*"), the genus "*gracile*," "*supranicum*," and "*britannicum*" (Böse 1973).

In the thirteenth century, two major works, Frederick II's "*De arte venandi cum avibus*" and Albertus Magnus's "*De falconibus*," marked the peak of Latin falconry

Fig. 4.3 The "*Epistola Aquilae Symachi et Theodotionis ad Ptolomeum regem Aegypti de avibus nobilibus*" employs an epistolary format. The letter, presumably addressed Ptolemy I Soter, the king of Egypt, was sent by three wise men of Greek origin: Aquila, Symachus, and Theodotion. Illustration of Ptolemy with a falcon, a miniature from folio 085v of "*Der naturen bloeme*" by Jacob van Maerlant (ca. 134–50; Credit: Wikipedia)

treatises (1200–1280; Stuhmiller 2015). Albertus Magnus, writing in the encyclopedist tradition, drew heavily from earlier writers like Frederick II and William the Falconer of King Roger. However, he introduced a new scientific approach, using morphological and behavioral traits to classify falcons (Oggins 1993; Tkacz 2009). Albertus distinguishes 17 falcon species, including the gyrfalcon, the lanner falcon, the peregrine falcon (in black, white, and red varieties), a blue-footed species, and the merlin ("*mirle*"), "*falco lapidarius*," and "*falco arborealis*" (Stadler 1916–1920). He defined these species by considering both their overall size, tail feather length, and wing-pointedness (Tkacz 2009).

Besides their undeniable historical importance, these treaties provide little useful ecological data. Their main contribution lies in establishing the foundation for classifying falcons in modern ornithology (see Sect. 5.1).

4.2.3 *The Past Gyrfalcon International Trade*

In the medieval avian commerce, the recurring loss or escape of raptors spurred a continual demand for replacements. This led to a meticulously organized and extensive trade in birds of prey throughout medieval Europe and Eurasia. Beginning in the late tenth and early eleventh centuries, raptors from Iceland and Norway travelled southward to Denmark, Germany, the Netherlands, Belgium, and England, eventually reaching the Mediterranean by the thirteenth century (see Fig. 4.4; Carus-Wilson 1962; Allsen 2006; Ægisson 2015; Prummel 2018). In Gottfried von Strasburg's "*Tristan*" (circa 1210), the protagonist mentions the cargo of a merchant ship from Norway, which included many peregrine falcons, merlins, and sparrowhawks (von Strassburg 1974). During the twelfth and thirteenth centuries, gyrfalcons captured by locals in Norway and Iceland became the preferred and most traded birds of prey across the continent (Allsen 2006; Prummel 2018). Falcons were generally valued at three times the price of goshawks (Carus-Wilson 1962). The most prized white gyrfalcons were taken in Iceland during autumn migration from Greenland (Vaughan 1982). In 1226, Emperor Frederick II granted Lübeck the title of Imperial Free City. As part of their imperial duty, the people of Lübeck had to annually provide their overlord with 12 gyrfalcons from Norway and Iceland, transported through Bergen port (Friedland 1984). In 1240, according to Imperial records, Frederick instructed Lübeck merchants to transport gyrfalcons to Italy (Huillard-Bréholles 1852–1861). The cost of 12 gyrfalcons was approximately 350 aurei, equivalent to the value of roughly 50 tons of cereal grains, which could sustain the livelihood of 100–200 people annually (Friedland 1984). Hanseatic League merchants dominated the trade of trained Scandinavian white gyrfalcons, which were highly valued in European courts as gifts (Fig. 4.4; Friedland 1984; Fritz 1993; Allsen 2006; Stuhmiller 2015). Additionally, gyrfalcons were brought to Muslim countries from Northern Europe through diplomatic gifts, Italian and Spanish merchants, and from Russia via Central Asia (Buquet 2021). In the late thirteenth century, falcon trading occurred in the Bruges market. By 1378, a Lübeck trading

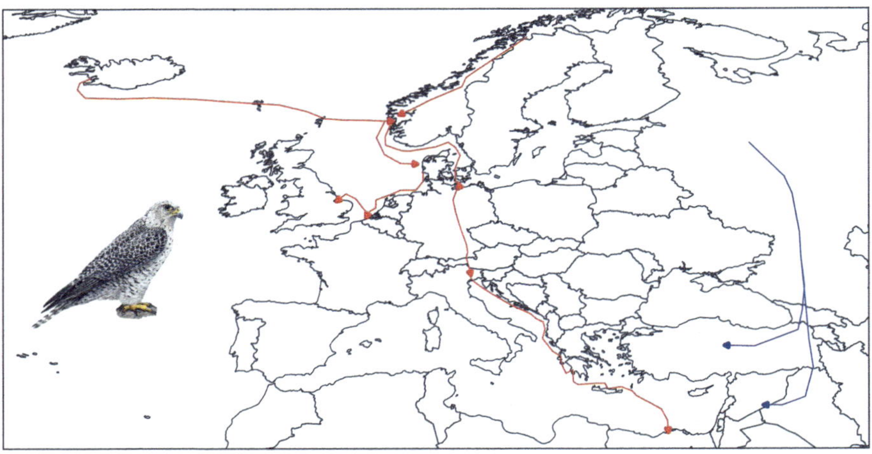

Fig. 4.4 During the Middle Ages, the primary trade routes for gyrfalcons extended from Iceland and Norway to Denmark or major Hanseatic League ports, such as Lübeck and Lynn (red lines). Falcons were also transported to Egypt via Venice, while the main routes to the Arab and Persian regions passed through Central Asia (blue lines; drawings courtesy of Marco Preziosi)

company was involved in selling northern falcons to Alexandria, Egypt, through Venice (Friedland 1984). In the fourteenth century, or even earlier, Swedish gyrfalcons extended their trade, reaching as far as Egypt (Fritz 1993). The historian Ismā'īl ibn 'Alī Abū l-Fidā' (1273–1331 CE) mentioned that beyond Ireland, there is the "*Island of gyrfalcons*" ("*Jazīrah al-sanāqir*"), referring to Iceland (Allsen 2006). In the Tārīh al-Manṣūrī by Ibn Naẓīf al-Ḥamawī (12th–13th c.), a Frankish ambassador named Messer Raimondo arrived in Egypt in 1232–1233 and bought a gyrfalcon for the Ayyubid sultan of Egypt, al-Malik al-Kāmil (d. 1238), at the high price of 300 gold ounces (Amari 1880). Despite the efficient market, Arabs and Persians learned about gyrfalcons through earlier exchanges with Central Asia (Buquet 2021).

Historical sources emphasize the extent of the gyrfalcon trade, but quantifying it is challenging. In 1351, Stephen Roserius of Montpellier obtained permission from Pope Clement VI to ship 25 gyrfalcons or falcons to Mamluk lands (Carr 2015). Another document 1327 mentions a significant purchase of 419 falcons, 107 of which were from a single Venetian merchant (Shedada 2012). A recent comprehensive study on bone remains from medieval Norway allocated only 0.17% to raptors (Walker et al. 2019). Within this group, the fact that falcons make up only 5% suggests that *Falco* species may have been less abundant (Fig. 4.5; Harrison et al. 2008). However, the presence of peregrines and gyrfalcons in Oslo and Bergen could indicate birds intended for trade with Europe (Lie 2018; Walker et al. 2019). It should be noted that the examined animal bone collections from the settlements mostly consist of kitchen and butchery waste (Schmölcke 2022). In addition, bone remains resulting from accidental deaths during transit must be considered (Harrison et al. 2008). The trade in gyrfalcons became quite lucrative, reaching its peak in

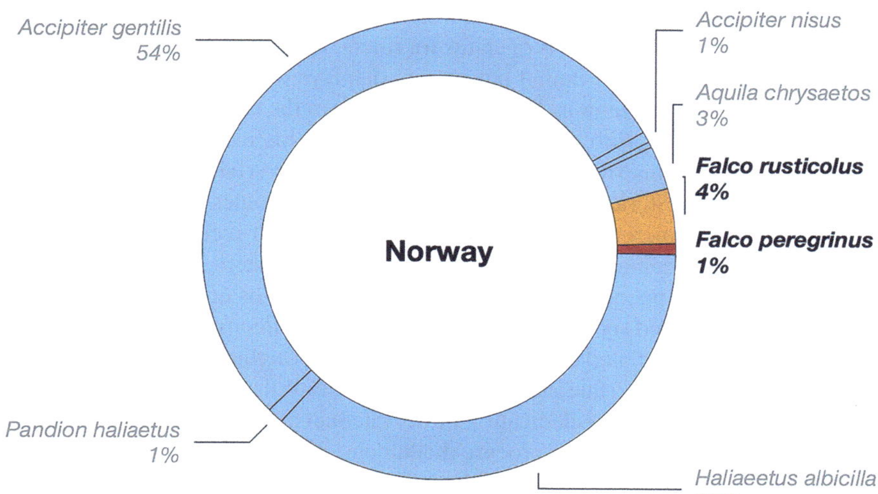

Fig. 4.5 Frequencies of raptors in bone material analyzed from 54 medieval settlement sites in Norway. (Data from: Walker et al. 2019)

1764, with over 200 birds exported per year, until it came to an end in 1810 (Petersen 1998). Lie (2018) suggests that exporting falconry birds from Norway may have reduced the abundance of *Falco* species in the Norwegian archaeological record.

4.3 Low Middle Age (11th–15th c. CE)

Most of the literature from this time originates from falconry treatises and elaborate courtly poetry, which often borrowed terminologies from falconry (see Sect. 4.2.1). However, until the eleventh century, the primary sources of information on falcons were falconry treatises written in the Eastern Islām (cf. Sect. 4.2.1; Akasoy and Georges 2005). In Europe during the same period, falcons were still included in funerary contexts, as evidenced by a female kestrel at Viborg Søndersø in Denmark (1016–1045 CE; Gotfredsen 2014). Arabic and Persian writings, on the other hand, focused on local species such as large desert falcons, which they could distinguish to some extent based on morphological differences. For example, the "*Kitāb al-kāfī fī al-bayzarah*" notes that the smaller male lanner falcon, often mistaken for the saker falcon, is found in Syria and the north-western Arabian Peninsula (Shehada 2012). The discovery of lanner falcon bone remains from the eleventh century inside the Great Mosque in ar-Rāfiqa, Syria, corroborates these observations (Becker 2005).

The rise and rapid expansion of falcon trading within the Christian world created a lucrative market and spurred an increase in the production of written works and art (cf. Sect. 4.2.3). During the Comnenian dynasty's rule over the Byzantine Empire (from 1057), falcons from Iberia were highly esteemed for their quality (Koukoules 1932). Additionally, large falcons were introduced to Europe by Crusaders returning from Palestine and from the Maghreb to France (Viré 1966; Attiya 1999). Falcons quickly became symbols of status for rulers, not only in death (as evidenced by their inclusion in burial items) but also as displays of power during their lifetimes. For example, the famous Bayeux Tapestry from the eleventh century depicts King Harald mounted with a hawk (not a falcon) on his left wrist (Yapp 1987). Kings procured falcons from distant regions, including gyrfalcons, and also from within their own territories if they were along migratory routes or in remote areas. The Palaiologoi emperors of Constantinople (11–12th c.) received falcons from Thessaly and Macedonia, common kestrels from southwestern Bulgaria, and hobbies from Thrace (Appendix; Koukoules 1932; Sidiropoulos et al. 2017). In 1245, King Henry III received six "*gerfalcons*" and six "*gentle falcons*" from King Haakon of Norway, who appreciated the scarlet cloth sent from England and frequently sent Henry prized northern gyrfalcons from Iceland (Carus-Wilson 1962). Haakon's letter to Henry emphasized the dedication of his falconers, who endured harsh conditions for two years in Iceland to procure the falcons, and expressed hope that Henry would value them as highly as previous monarchs had (Carus-Wilson 1962). In his book "*Il Milione*" (1298), Marco Polo mentioned several falcons used for falconry, including saker and lanner falcons (Bertolucci Pizzorusso 1975). He also referred an island in the north (Iceland) from which the Great Khan obtains his gyrfalcons

(Vaughan 1982). Another traveller, Giraldus Cambrensis (c. 1146–c. 1220), in his book "*The Itinerary of Archbishop Baldwin Through Wales*," made notable observations for his time. He recognized that female falcons are larger than males and noted that despite breeding efforts, the falcon population did not increase, which may represent one of the earliest recorded observations of natural population regulation.

During the Low Middle Age, falcons became a prominent poetic archetype in courtly literature (Abeele 1990). Drawing from practical falconry knowledge and experience, they were associated with various human traits. Traits such as fierceness, courage, cowardice, and jealousy were often compared to the behaviors exhibited by falcons during their hunting activities. For example, the minstrel and poet Badouin de Condé (d. 1272) made a disparaging comparison between the lanner falcon and other raptors that were more adept at pursuing large prey (Abeele 1990). Unlike the powerful gyrfalcon, saker, and peregrine falcon, favored by noble rulers, the lanner falcon and other smaller or less capable falcons were considered inferior ("*ignobilium genera*") in medieval times (Smets 2002). High-status ladies frequently participated in hawking parties, often using lanner falcon males, known as "*lannerets*," and merlins (Oggins 1989). François Rabelais (1494–1553) described this practice, noting: "*Les dames montées sur belles hacquenées portaitent chascune ou un épervier, ou un lanneret, ou un émérillon.*" Falconry imagery, associated with courtly love, often symbolized the pursuit of affection (Fig. 4.6; Standley

Fig. 4.6 The renowned "*Codex Manesse*" manuscript (c. 1304–1340), an anthology of poems and songs, features several illustrations of falcons. These images offer evidence that both men and women of the nobility participated in falconry. On the right, the poet Werner von Teufen is depicted embracing a lady who holds a falcon or hawk. (Permanent link: https://doi.org/10.11588/diglit.2222#0009 and https://doi.org/10.11588/diglit.2222#0133)

2008). In the fourteenth-century "*Codex Manesse*," a couple on horseback is depicted, with the man amorously embracing a lady who holds a falcon or a hawk on her gloved hand (Fig. 4.6). In medieval German poetry, falcons are frequently associated with hesitant ladies and poets or with those who seem to lose both noble falcons and their beloved partners (Dalby 1965). By the 1400s, the poetic falcon had largely disappeared from European literary texts, likely reflecting the decline of its associated literary genre, "*le chanson de geste*" (Leonardi and Sutton 2020).

4.3.1 De Arte Venandi Cum Avibus

In the Middle Ages, falcons were symbols of both Muslim emirs and Christian princes (Masseti 2009). Falconry served as a bridge between the elite of Western Christianity and the Islamic world, influenced by Persian traditions, and it spread from the courts of the Umayyad and Abbasid dynasties to Christian Europe (Martinez Enamorado 2011). When falconry gained popularity in the Kingdom of Sicily, a translation school in Palermo compiled information from Arabic sources (Battaglia 1977). Around 1240–1241, Emperor Frederick II of Hohenstaufen (1194–1250) utilized these translations, known as the "*Falkenbuch.*" He commissioned Latin translations of works by notable Arab authors such as *Moamin* and *Adham*, as well as the Persian *Ghitrīf*, specifically for the Hohenstaufen dynasty (Battaglia 1977; Allen 1980; Oggins 1993; Martinez Enamorado 2011; see Sect. 4.2.1). Theodore of Antioch, a Syrian doctor and scientist, translated *Moanin*'s work, and the Frederick himself contributed several corrections to the Latin texts (Trombetti Budriesi 2000). Frederick's extensive knowledge of falcons and other predators was gained through direct experience. In 1231, he visited Ravenna "*with many animals unknown to Italy: elephants, dromedaries, camels, panthers, gerfalcons, lions, leopards, white falcons, and bearded owls*" (Haskins 1971).

Frederick II imported falcons from the Middle East, but records indicate that he also dispatched his falconers, including Stefano Mustaccius, to collect birds in Sicily, nearby islands, and Malta (Bresc 1980). The renowned treaty "*De Arte Venandi Cum Avibus*" (Pal. lat. 1071; Biblioteca Apostolica Vaticana; permanent link http://digi.vatlib.it/view/bav_pal_lat_1071/0001), attributed to Frederick, is a comprehensive Latin compilation on falconry that has been extensively cited in later treatises (Oggins 1993). It does not simply quote ancient texts but provides a critical review and incorporates insights from field observations.

The text provides a thorough discussion of the larger falcon species used in falconry, specifically "*Girofalcones*," "*falcones sacrii*," "*falcones peregrini*," and "*falcones lanerii*" (or "*laynerii*"; Yapp 1982; Trombetti Budriesi 2000). Additionally, it accurately describes the common kestrel's ability to hover while hunting (Trombetti Budriesi 2000). Interestingly, based on plumage patterns, the author distinguishes between the migratory peregrine falcon ("*falco gentilis peregrinus*") and the resident subspecies in central Europe ("*falco gentilis absolute*"; Trombetti Budriesi 2000). Besides describing their morphological traits, the book also explains the geographic origins of each species and their adaptation to different climate zones (Fig. 4.7). These zones, derived from Ptolemaic concepts, are clearly defined in Al-Idrīsī's cartographic work known as "*The Book of Roger*," dating back to 1154 CE (Fig. 4.7). Frederick II was well aware that the preferred breeding grounds for gyrfalcons were the sea cliffs of Norway, Iceland, and Greenland ("*Galandia*") near the North Pole ("*polum arcticum*"; Trombetti Budriesi 2000; Allsen 2006; Buquet 2021). He notes that gyrfalcons come from Iceland ("*Yslandia*") in the seventh climatic zone ("*septimi climatis*") and do not extend beyond this to the sixth, fifth, and fourth zones. Similarly, the migratory peregrine falcons (*F. peregrinus calidus*) originate from the seventh zone, with resident falcons living in the southern parts of that zone. The information on saker and lanner falcons is inconsistent and confused. The text uses the adjective "*vero*," meaning "*true*," for both "*sacrii*" and "*lanerii*" (Yapp 1982). Saker falcons nest in the seventh zone and south of it, as seen in Bulgaria, and they have also been found in areas never previously occupied, such as Britannia (= Anglia; Trombetti Budriesi 2000). Regarding the distribution of the lanner falcon, Frederick II claims that "*lanners breed in all climates*" (Trombetti Budriesi 2000). Due to this incorrect assessment of the breeding range, Yapp (1983)

Fig. 4.7 The seven climatic zones (highlighted in red) as defined by the geographer Muhammad Al-Idrīsī in "*The Book of Roger*". (Credit: Wikipedia)

suggests that Frederick II regarded the names "*layner*" and "*sacer*" as merely variations of the peregrine falcon.

The illustrations in "*De Arte Venandi cum Avibus*" are not detailed by medieval standards, possibly because Frederick II, who passed away before its publication, could not personally supervise it. Except for the clear depiction of gyrfalcons on folio53, falcons are shown in a highly stylized manner. The definitive identification of gyrfalcons relies more on their whitish coloration than on specific plumage patterns or morphological traits. Similarly, the illustration at the foot of folio56 of the chapter "*De laneriis*" shows five birds, none of which have the diagnostic rufous crown of the modern lanner falcon (Yapp 1983). The illustration on folio67, depicting a large falcon hunting small mammals and lizards, also does not resemble a lanner falcon (Yapp 1983). Conversely, a falcon with an almost chestnut crown portrayed on folio55 does resemble a lanner falcon, but the text description identifies it as a peregrine falcon (Yapp 1983).

4.3.2 Falcon Catalogues for Their Management

In markets where goods are traded, efficient management is crucial, including handling payments and associated costs. For falcons, this was particularly challenging due to the need for full-time professional care, leading to significant expenses. Additionally, falcons were often given as gifts to enhance diplomatic relationships between rulers. Therefore, besides managing them, there was a constant need to maintain falcon stocks by capturing them in the wild. This required meticulous tracking of costs and revenues, along with management decisions made by the rulers. These records offer a unique opportunity to quantify the falcons involved in this market. Unfortunately, the administrative staff were more attentive about to financial details than to the accurate naming of the falcons. Moreover, there has been long-standing confusion about the proper identification and naming of falcons. Despite these limitations, these sources can provide reliable data about the historical distribution of falcons.

From the administrative documents of Frederick II's reign, we learn about the capture of falcons in Sicily and the surrounding islands, as well as in Apulia and Malta (Huillard-Bréholles 1852–1861; Bresc 1980). These were presumably peregrine falcons, both native and migratory, the latter captured on small islands or in Malta (Fenech 2010). It is also plausible that lanner falcons were caught both in Sicily, where they have their stronghold, and Apulia (Leonardi and Sutton 2020). This fact was further confirmed three centuries later by Charles d'Accursia (d'Accursia 1598). In many instances, the names "*faucons dit laniers*," "*pieds bleus*," "*blāfuss*," or "*blue legs*" cannot be attributed to the lanner falcon species (cf. Sect. 4.4.2). Nevertheless, analyzing and interpreting these names can still provide valuable insights for more precise species identification. The "*Kitāb fī tafṣīl al-zamān wa-maṣāliḥ al-abdān*" ("*Calendar of Córdoba*"), likely written by ᶜArīb bin Saᶜīd (d. 980), a Muslim physician, and Rabīᶜ bin Zayd (fl. second half tenth century), includes discussions on meteorological and astronomical topics, such as astronomical tables, health and hygiene, agriculture, and calendars (Fradejas Rueda 2008; Forcada 2011). In the months of January, March, and April, a falcon breeding on sea cliffs called "*shaḏḫāniqat al-baḥriyya*" is mentioned, while other nonbreeding falcons are called "*shaḏḫāniqat al-libliyya*" (Viré 1965; Fradejas Rueda 2008). Fradejas Rueda (2008) suggests that "*shaḏḫāniqat al-baḥriyya*" specifically refers to the subspecies of peregrine falcon commonly known as the barbary falcon (*F. p. pelegrinoides*). Thus, the Arabism in the Spanish language, "*baharí*" (or "*bafarí*" in Portuguese), has been in use since at least 1250 (Fradejas Rueda 2008). The practice of distinguishing falcons based on the most favorable period for capturing them was already well described by Frederick II. Additionally, it was widely recognized that certain areas were particularly suitable for capturing falcons, likely due to the presence of migratory birds. The Alpine region around Chur was considered one of these, as officials there collected a fee from people who caught falcons (Goldberg 2020).

There are numerous European catalogues in which captured falcons have been registered, purchased, and sold (Oggins 2004). The administrative records (1300–1315) of the Counts of Artois and Flanders at Hesdin (northern France) note 17 falcons, including 10 common kestrels (and/or Eurasian hobbies), 6 peregrine falcons, and 1 "*faucon dit lanier*" (Duceppe-Lamarre 2000). In Haarlemmerhout (The Netherlands), Jean de Blois reported in 1364 that there were 14 peregrine subadult females, 23 sub-adult males, and 4 lanner falcons (Govaerts 2024; see also Sect. 4.4.2). Expanding the time frame between 1354 and 1390, a total of 64 peregrine falcons, 7 lanner falcons, 1 saker falcon, 1 gyrfalcon, and 1 merlin are listed as captured (Govaerts 2024). Excluding the unreliable data about the lanner falcon, fossil evidence in the area shows sparrowhawks, common buzzards (*Buteo buteo*), long-eared owls (*Asio otus*), and an unidentified member of the genus *Falco* (Govaerts 2024). It is important to note that many falcons at the royal court were either gifts or purchased. During Edward I's reign in England (1277–1307), gyrfalcons outnumbered peregrine falcons by 26%, not counting the 25% listed as "*lanner falcon*" (Oggins 2004).

The "*Great Marienburg Tressler Book*," covering the years 1399–1409, details expenditures by the Teutonic Order and the Duchy of Prussia, specifically the Marienburg convent and the Grand Master, on the purchase and sale of falcons and other raptors (Heckmann and Heckmann 2017). These records include references to peregrine and gyrfalcons. According to Heckmann and Heckmann (2017), there may be a correlation between the frequency of captures and climatic fluctuations in the Baltic Sea region (Figs. 4.8 and 4.9). Today, the gyrfalcon is considered a vagrant in the same area. Therefore, while an average of one individual per year in the past might seem plausible, it is clear that there were also instances of purchased specimens (Fig. 4.8). The frequencies of peregrine falcons appear to be predominantly correlated with historical weather events in the Baltic Sea area (Fig. 4.9).

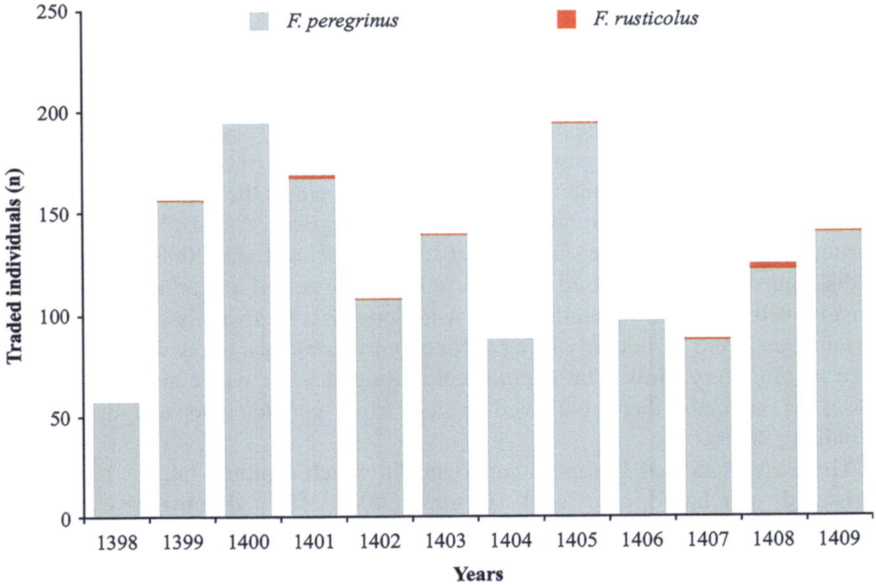

Fig. 4.8 Peregrine (n = 1563) and gyrfalcons (n = 11) recorded in the "*Great Marienburg Tressler Book*" from 1399 to 1409. (Data from Heckmann and Heckmann 2017)

4.4 Western Palearctic Falcons

The Middle Ages are critical for understanding the relationship between humans and falcons. Historical sources reveal a robust trade and significant exploitation of both native and migratory falcon populations (see Sects. 4.2.3 and 4.3.2). However, this apparent abundance was largely shaped by market demands. This influence is evident when examining the percentage frequencies of raptors across various subperiods of the Middle Ages (Fig. 4.10). As a result, key falconry species, such as the gyrfalcon, are overrepresented, while species not used in falconry, like Eleonora's and the red-footed falcon, are entirely absent (Appendix). Disentangling these data to ascertain the true field frequencies proves challenging and unreliable, in specific cases (see Sect. 4.3.2). Among the four sources discussed at the beginning of this chapter, only fossils and catalogues offer a reliable account. Additionally, past distributions of falcons in the Palearctic should be interpreted with caution due to inconsistent taxonomic alignment and potential inaccuracies from falconry community data. This caution is warranted because modern ornithology, which developed later, has sometimes accepted these historical sources as accurate (see Chap. 5). Although nineteenth-century experts like Schlegel (1844) sought to align their data with that of Pierre Belon (1555), some of that taxonomy remains in use today.

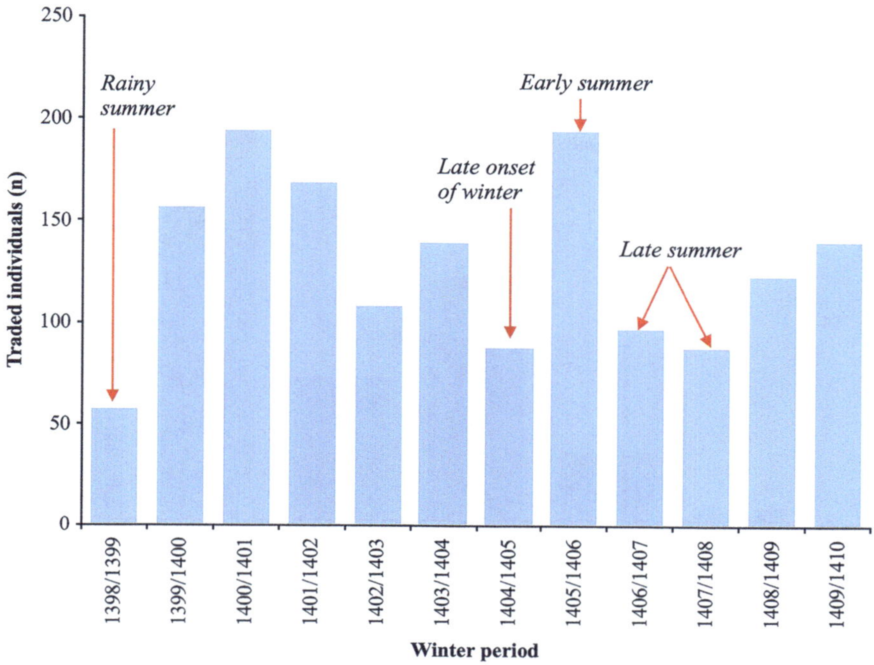

Fig. 4.9 Peregrine falcons recorded in the "*Great Marienburg Tressler Book*" (1399–1409) in relation to specific weather events in Central Europe, the Baltic Sea, and the Kola Peninsula. (Data from Heckmann and Heckmann 2017)

4.4.1 Small Falcons

Traces of small falcons, whether fossil or written, are scarce and fragmented outside the context of falconry (Appendix). As Negro et al. (2020) suggest, small falcons may have been too common in the past to be considered noteworthy. This is particularly true for species such as the lesser kestrel, which adapted early to nesting within human settlements (Negro et al. 2020). Before the nineteenth century, lesser and common kestrels were not distinguished as separate species, though some authors identified them as "*urban*" and "*wood*" kestrels (Negro et al. 2020). Despite the lack of fossil evidence, it is known that lesser kestrels began nesting in palaces in the fifth century and continued this behavior through the seventeenth century, particularly in Andalucía, Southern Spain (Appendix; Negro et al. 2020).

The common kestrel, while regarded as a low-value falcon, was still used in falconry by commoners (Tyrberg 2002). However, it is important to note that because common kestrels frequently nest on man-made structures, discovering their bones does not necessarily indicate their use in falconry. Additionally, the common kestrel was the most frequently encountered falcon species during this historical period (see Appendix). The name originates from the Middle French "*cresserelle*" or

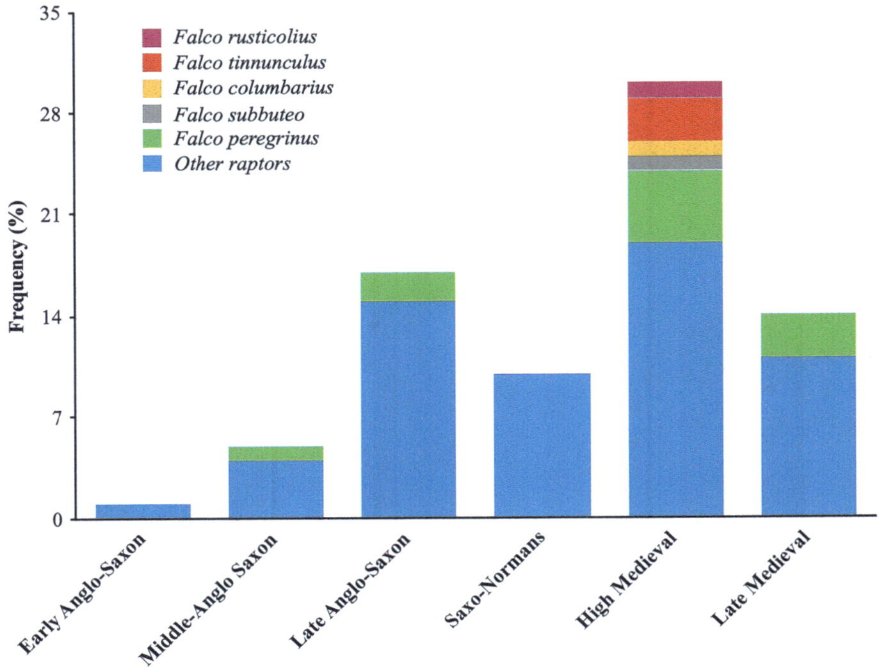

Fig. 4.10 Frequencies of falcon species in raptor bone assemblages across different sub-periods of the Middle Ages. (Data from Poole 2018): Middle-Anglo Saxon (650–850 CE), Late Anglo-Saxon (850–1066 CE), Saxo-Normans (1066–middle 12th c.), High Medieval (middle 12th–middle 14th c.), and Late Medieval (middle 14th–15th c. CE)

"*crecerelle*," which alludes to its shrill vocalizations, similar to the Roman use of "*tinnunculus*" (see Sect. 3.6.1; Myers 2022).

The only fossil evidence of the red-footed falcon from this historical period was discovered at Obishir-5 in Kyrgyzstan (Serdyuk et al. 2023). This specimen is likely a sub-adult from a population that has since become extinct, possibly due to a prolonged expansion process during the late Holocene (Serdyuk et al. 2023). The species has already experienced a documented longitudinal expansion at the end of the Pleistocene (see Chap. 1), which may have continued or been repeated during the Holocene.

Despite being named after a historical figure, there are no records of Eleonora's falcon. Eleonora d'Arborea, Sardinia's most celebrated heroine, enacted a prohibition against removing goshawks and falcons from their nests in the "*Carte de Logu*" (1392; Lupinu 2022). However, these references were not to Eleonora's falcons specifically; they pertained to large falcons, such as the peregrine falcon, which she used in falconry (Landsborough Thomson 1976).

The merlin was one of the most frequently used small falcons in medieval falconry, alongside the common kestrel. It is well-known, with its name tracing back to Proto-Germanic "*schmerl*" and the Old Norse "*smyrill*" (Fonteijne 1993; Myers 2022). The term evolved from "*émerillon*" in Old French, through "*meriliun*" in Anglo-French, then to "*merlion*" in Middle English (1350–1400), and finally to "*merlin*" (Fonteijne 1993; Myers 2022). Fossil evidence points to an active trade in this species, with recorded trade locations in Northern Europe, including England and Scandinavian countries, but no southern locations as seen in earlier periods (see Sect. 3.7.1 and Appendix).

Fossil evidence for Eurasian hobbies is limited, mainly relating to trade and capture sites (e.g., Didymoteicho, Thracia, Greece; Appendix). Despite their wide distribution and use in falconry, the name of this bird originates from the falconry terminology (Myers 2022). The name evolved from "*hoberel*" (1195), to "*hobel*," and later "*hobereau*" from 1377 (Myers 2022).

4.4.2 Large Falcons

Large falcons were primarily targeted for capture, trade, and use in falconry, in contrast to smaller species. However, fossil evidence is limited for some species, while others, like the gyrfalcon, are disproportionately represented. Moreover, a comprehensive analysis is necessary for accurately revising historical sources.

The lanner falcon presents significant challenges, likely due to its population being fragmented into small, localized demes (Leonardi 2015). This fragmentation has caused confusion with other large falcons and led to inconsistent naming (see Leonardi and Sutton 2020 for a review). Fossil remains of this species are found in only two locations: Hungary and Syria (Appendix). Accordingly, there is a notable scarcity of fossils from this species dating back from the Pleistocene (see Chaps. 1 and 2). In Hungary, the remains are likely from a falcon trained for falconry, possibly brought during one of King Sigismund's military expeditions (Gál 2015). The remains discovered in Syria are from an individual captured in the eleventh century and kept in the Great Mosque in the center of ar-Rāfiqa (Becker 2005). Despite the species being involved in numerous commercial transactions in Europe under various names, no fossil evidence confirms its historical abundance. This confusion stems from the fact that, during the Middle Ages, the name "*lanier*" or "*blue legs*" predominantly referred to what is now recognized as *Falco biarmicus* (Lindner 1962).

While historians continue to support this view, ornithologists challenge it (Lindner 1962; Kumerloeve 1963; Mayaud 1963; Leonardi and Sutton 2020; Govaerts 2024). The debate began with Lindner (1962), who used falconry management catalogues to compare the historical and current distribution of species (Fig. 4.11). Lindner proposed that lanner falcons inhabited Central Europe (Germany, France, and Switzerland) from the thirteenth to sixteenth centuries. He also noted that lanner falcons were among the most commonly used birds of prey in high medieval and early modern Europe (Lindner 1962). However, recent research based on catalogues has shifted their northern limit to the Netherlands, where, in favorable years, lanner falcons have sometimes outnumbered peregrines (Govaerts

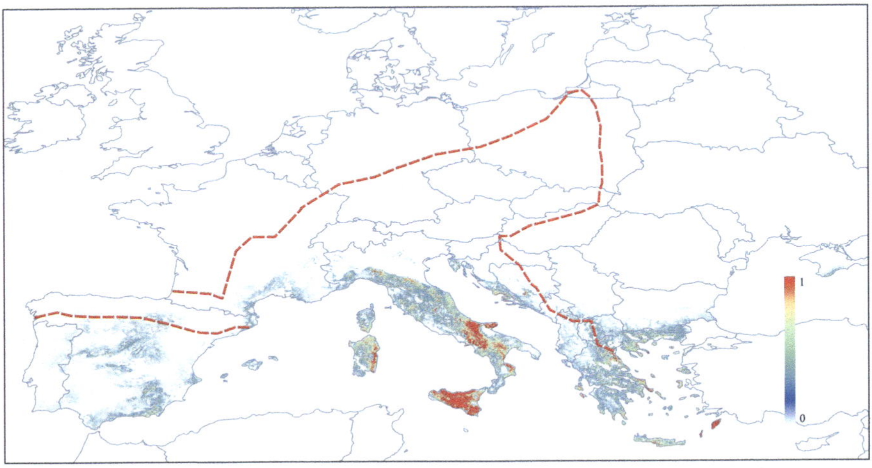

Fig. 4.11 The dashed red line indicates the presumed distribution of the lanner falcon in Europe during the Middle Ages. (Redrawn from Lindner 1962). Additionally, the figure illustrates the potential distribution of the species, with probability values ranging from 0 (minimum) to 1 (maximum), based on the environmental conditions of the lanner falcon's nesting sites

2024). The primary counterargument is that the term "*blue legs*" is too imprecise, encompassing various species, which makes leg color an unreliable criterion for identifying falcons (Mayaud 1963; Dalby 1965). Additionally, some suggest that these names might simply refer to the more common peregrine falcon (Yapp 1983). Lindner argued that confusion over the term "*blue legs*" arose because the lanner falcon species retreated to South and Southeast Europe (Lindner 1962). He attributed the species' expansion to the favorable climate during the Medieval Warm Period and deforestation, while contraction was linked to human persecution (Govaerts 2024). If such expansions occurred, similar to those of the gyrfalcon during the Last Glacial Maximum or the red-footed falcon, they would likely involve longer timescales and more significant climate changes (see Chaps. 1 and 2). Furthermore, the chance of finding fossil remains in both natural areas and human settlements should be high, as evidenced by the peregrine falcon (see Appendix). The complete absence of lanner falcon fossils in Europe reflects the rarity and fragmented populations even in earlier periods (see Chaps. 1 and 2). This rarity is ultimately attributed to the lanner falcon's specialized climate requirements, which contrast with the broad niche of the peregrine falcon and the intermediate adaptability of the gyrfalcon (Fig. 4.11; Popescu et al. 2024).

Even for a species like the saker, which has a specialized climate niche, fossils are rare and primarily linked to falconry (Appendix). The species Frederick II referred to as "*sacrii*" is difficult to classify in modern taxonomy, but it was undoubtedly present in European courts as both adults and *juveniles*. In this context, it is highly likely that these individuals were classified as "*lanier*" and/or peregrine falcons. However, administrative documents from Malta in 1468 mention that both male and female sakers were captured on the island (Fiorini 2001). Interestingly, later documents from the seventeenth century distinguish between "*lanier de Russie*" (likely the saker falcon) and the "*lanier de Sicilie*" (probably the lanner falcon; see Chap. 5; Fenech 2010). Similar to the lanner falcon, Lindner (1962) suggested that the saker falcon bred continuously in central Europe during the Middle Ages, but Baumgart (1980) confuted this assertion. Fossil remains are mainly found in Eastern Europe, aligning with the likelihood of capture in areas with higher relative abundance (Appendix).

Information about the gyrfalcon primarily comes from records of its trade during the High Middle Ages (see Sect. 4.2.3 and Fig. 4.4). The name originates from the Proto-Germanic word "*giri*" (=greed), which evolved through Old French as "*gerfaucon*" and Old Norse as "*geirfalki*" before reaching other European languages (Jobling 2011; Myers 2022). Frederick II referred to them as "*girofalcones*" (Trombetti Budriesi 2000). During the Middle Ages, the main strongholds from which gyrfalcon were sourced for continental Europe and the Middle East were the northern areas of Norway, Iceland, and Greenland (Fig. 4.4). These historical breeding areas were similar to the current ones, shaped by the retreat of the perennial ice after the Last Glacial Maximum period (Chaps. 1 and 2). During the Medieval Warm period, rising temperatures allowed the species to temporarily colonize areas in Greenland that are normally covered in ice (Appendix; Moseley et al. 2019).

Even in the Middle Ages, the peregrine falcon was the most common large falcon species in the Western Palearctic (Appendix). This is supported by numerous fossilized remains, trade records, and administrative and literary sources (see Sect. 4.3.2). Its wide distribution is also evidenced by its presence as a fossil from periods before the peak of trade in the High Middle Ages (Fig. 4.10). Albertus Magnus was likely the first to use the name "*peregrine*" (meaning wanderer), referring to the migratory habits of this species (Myers 2022). The peregrine falcon was the main species used in falconry, particularly the larger females, which were praised in the literature but poorly depicted in illustrations (Fig. 4.12; Yapp 1979, 1983, 1987).

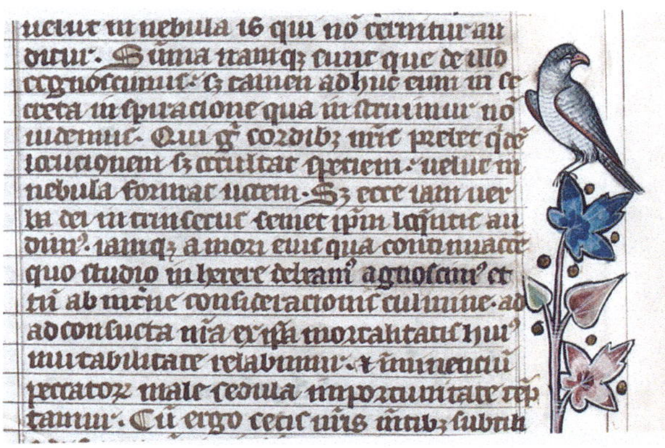

Fig. 4.12 Decorative peregrine falcon depicted on page f182.v of the MS 112, "*Selected illuminations from Gregory the Great's Moralia in Iob*". (Credit: by permission of The Master and Fellows of Emmanuel College, Cambridge)

4.5 Conclusions

Despite the wealth of information about falcons from various sources, reliable data remains limited and should be approached with caution. The Middle Ages, spanning a long period, witnessed significant changes in the relationship between humans and falcons. Falconry reached its zenith and then gradually declined. During this time, widespread interest in falcons led to numerous publications, both specialized and literary, which compiled contemporary knowledge of falcon species.

Several key points can be highlighted:

- The importation of exotic falcon species into different regions and ongoing misidentification have contributed to confusion regarding their historical distribution.
- The naming of falcon species became more standardized due to the development of technical terms in falconry.
- Species that were either common or of little importance in falconry have left scant evidence in historical sources.
- The increasing number of captures in the wild for some species, such as the gyrfalcon, may have had a significant negative impact on nesting and migrating populations.
- The knowledge acquired during this period was crucial in establishing the foundations for classifying falcons within the emerging field of ornithology.

Appendix: Fossil Remains (*f*), Artifacts (*a*), and Literature Records (*r*) of Western Palearctic Falcons from the Middle Ages (7th–15th c. CE)

Species	Dates	Location	References
F. naumanni	–	–	–
F. tinnunculus[(f)]	8th–9th and 11th century	Starigard/Oldenburg, Germany	Teegen (2018)
F. tinnunculus[(f)]	9th–12th century	Arkona sanctuary, Germany	Teegen (2018)
F. tinnunculus[(f)]	Early Middle Ages	Menzlin, Germany	Teegen (2018)
F. tinnunculus[(f)]	Early 14th–18th century	Koło castle, Poland	Bochenski et al. (2018)
F. tinnunculus[(f)]	8th–9th–13th century	Santok, Poland	Bochenski et al. (2018)
F. tinnunculus[(f)]	11th–early 13th century	Kałdus, Poland	Bochenski et al. (2018)

(continued)

Species	Dates	Location	References
F. tinnunculus(f)	8th–mid-13th century	Legnica, Poland	Bochenski et al. (2018)
F. tinnunculus(f)	9th–10/11th century	Myślibórz, Poland	Bochenski et al. (2018)
F. tinnunculus(f)	10th–13th centuries	Wrocław-Ostrów Tumski, Poland	Bochenski et al. (2018)
F. tinnunculus(f)	13th–18th century	Kraków, Poland	Bochenski et al. (2018)
F. tinnunculus(f)	Late Saxon (850–1066 CE)	Vicarage Garden, Brixworth, UK	Poole (2018)
F. tinnunculus(f)	1262–1321 CE	Dudley Castle, UK	Poole (2018)
F. tinnunculus(f)	13th–early 14th century	Thrislington, UK	Poole (2018)
F. tinnunculus(f)	13th–14th century	Thrislington Village, UK	Poole (2018)
F. tinnunculus(f)	1450–1550 CE	West Cotton, UK	Poole (2018)
F. tinnunculus(f)	Late 14th–16th century	St Gregory's Priory, Canterbury, UK	Poole (2018)
F. tinnunculus(f)	1016–1045 CE	Viborg Sødersø, Denmark	Gotfredsen (2014)
F. tinnunculus(f)	10th–11th century	Stará Boleslav, Czech Republic	Mlíkovský (2004)
F. tinnunculus(f)	13th–16th century	Felsőnyék-Várhegy, Hungary	Jánossy (1985)
F. tinnunculus(f)	15th–16th century	Székelykeresztúr-Udvarház, HU	Gál (2008b)
F. tinnunculus(f)	11–14th century CE	Hrodna, Belarus	Gorobets and Kovalchuk (2017)
F. tinnunculus(f)	11–14th century CE	Leeuwarden, The Netherlands	Prummel and Heinrich (2005)
F. tinnunculus(f)	ca. 1000–1500 CE	Groningen, The Netherlands	Prummel and Küchelmann (2022)
F. tinnunculus(f)	13th–14th c. CE	La Charité-sur-Loire, France	Audoin-Rouzeau (1986)
F. tinnunculus(f)	8th c. CE (falconry)	Wellin, Belgium	Prummel (2018)
F. tinnunculus(f)	c. 1157–c. 1227	Launceston Castle, UK	Albarella and Davis (1994)
F. tinnunculus(r)	11th–12th c. CE	Mount Meleniko, SW Bulgaria	Koukoules (1932)
F. tinnunculus(f)	15th–16th c. CE	Beja, Portugal	Moreno-García and Pimenta (2010)
F. tinnunculus(f)	Early Middle Age	Kałdus, Poland	Makowiecki and Makowiecka (2013)

Species	Dates	Location	References
F. vespertinus[f]	Middle Age	Obishir-5, Kyrgyzstan	Serdyuk et al. (2023)
F. eleonorae	–	–	–
F. columbarius[f]	14th century (falconry)	Vilnius Lower Castle, Lithuania	Girininkas and Daugnora (2018)
F. columbarius[f]	8th–9th century (falconry)	Starigard/Oldenburg, Germany	Teegen (2018)
F. columbarius[f]	8th–9th century (falconry)	Groß Strömkendorf, Germany	Teegen (2018)
F. columbarius[f]	Late 12th–13th century (trade)	Copt Hay, UK	Poole (2018)
F. columbarius[f]	8th century	Staraya Ladoga, Russia	Shaymuratova et al. (2019)
F. columbarius[f]	13th–15th century CE	Tver, Russia	Zinoviev (2021)
F. columbarius[f]	7th–8th century CE	Vendlas hög, Sweden	Vretemark (2018)
F. subbuteo[f]	15th–18th century	Tykocin, Poland	Bochenski et al. (2018)
F. subbuteo[f]	12th–13th century	Middleton Stoney Castle, UK	Poole (2018)
F. subbuteo[f]	10th–12th century	Budeč, Czech Republic	Peške (1993)
F. subbuteo[f]	10–13th century CE	Novgorod, Russia	Hamilton-Dyer (2002)
F. subbuteo[f]	11th–14th century CE	Hrodna, Belarus	Gorobets and Kovalchuk (2017)
F. subbuteo[r]	11th–12th century CE	Didymoteicho, Thracia, Greece	Koukoules (1932)
F. biarmicus[f]	14th century CE	Budavár-Teleki Palota, Hungary	Gál (2008a)
F. biarmicus[f]	11th century CE	ar-Rāfiqa, Syria	Becker (2005)
F. cherrug[f]	14th century (falconry)	Vilnius Lower Castle, Lithuania	Girininkas and Daugnora (2018)
F. cherrug[f]	10th–15th century	Middle Volga, Russia	Galimova et al. (2014)
F. cherrug[r]	13th century	Duganovo, Bulgaria	Arabadhiev (1962)
F. cherrug[f]	12th–13th century	Voin', Ukraine	Gorobets and Kovalchuk (2017)
F. cf. cherrug[f]	650–750 CE	Salme, Estonia	Peets et al. (2021)
F. cherrug[f]	10th–14th c. CE (falconry?)	Debrecen-Monostor-erdő, Hungary	Gál et al. (2020)
F. cherrug[f]	13th–14th c. CE	Tatarstan, Russia	Askeyev et al. (2013)
F. cherrug[f]	12th–13th c. CE	Elabuga, Tatarstan, Russia	Shaymuratova et al. (2023)
F. cherrug[f]	13th–15th c. CE	Maly Sundyr', Mari El, Russia	Shaymuratova et al. (2023)
F. cherrug[r]	1468 CE	Malta	Fiorini (2001)
F. rusticolus[f]	980–1450 CE (trade)	Gásir, Iceland	Harrison (2009)

Species	Dates	Location	References
F. rusticolus(f)	14th century (falconry)	Vilnius Lower Castle, Lithuania	Girininkas and Daugnora (2018)
F. rusticolus(f)	12th–14th century (falconry)	Winchester, UK	Poole (2018)
F. rusticolus(f)	1350–1510 CE (falconry)	Southampton, UK	Poole (2018)
F. rusticolus(r)	8th century (trade)	Kent, UK	Dobney and Jaques (2002)
F. rusticolus(f)	12th century (falconry)	Söderköping, Östergötland, Sweden	Vretemark (1997)
F. rusticolus(f)	11th–15th century (trade)	Bergen, Norway	Walker et al. (2019)
F. rusticolus(r)	c. 1380 CE (falconry)	La Rochelle, Oleron, France	Fradejas Lebrero and Rueda (2016)
F. rusticolus(f)	8th century CE	Vendel, Uppland, Sweden	Schmölcke (2022)
F. rusticolus(f)	11th–13th century CE	Eketorp on Öland, Sweden	Schmölcke (2022)
F. rusticolus(f)	9th–10th century CE	Söderby, Sweden	Schmölcke (2022)
F. rusticolus(f)	1025–1325 CE	Sandnes, Greenland	McGovern et al. (1996)
F. rusticolus(f)	1000–1500 CE	Narsaq, Greenland	McGovern et al. (1993)
F. rusticolus(f)	ca. 850–1300 CE	Kronprins Christian L., Greenland	Moseley et al. (2019)
F. rusticolus(f)	10th–early 13th CE	Bilyar, Republic of Tatarstan, Russia	Shaymuratova et al. (2023)
F. peregrinus(f)	10th–early 13th c. CE	Bilyar, Republic of Tatarstan, Russia	Shaymuratova et al. (2023)
F. peregrinus(f)	13th–14th c. CE	Uvek, Saratov region, Russia	Shaymuratova et al. (2023)
F. peregrinus(f)	12th–13th century CE	Voin', Ukraine	Gorobets and Kovalchuk (2017)
F. peregrinus(f)	10th century CE	Pliska, Bulgaria	Boev (2016)
F. peregrinus(f)	11th–15th century (trade)	Bergen, Norway	Walker et al. (2019)
F. peregrinus(f)	9th–10th century (falconry)	Starigard/Oldenburg, Germany	Teegen (2018)
F. peregrinus(f)	8th–9th century (falconry)	Groß Strömkendorf, Germany	Bleile (2018)
F. peregrinus(f)	8th–11th century (emporium)	Haithabu, Germany	Bleile (2018)
F. peregrinus(f)	11th–13th century (city)	Schleswig (Schild), Germany	Bleile (2018)
F. peregrinus(f)	Middle Ages	Hedehusum, Germany	Sten and Vretemark (1988)
F. peregrinus(f)	13th–15th c.	Burg Nürnberg, Germany	Prummel (2013)
F. peregrinus(f)	c. 1200 CE	Stendal, Germany	Prummel (2013)
F. peregrinus(f)	Late Saxon (850–1066 CE)	Ipswich, UK	Poole (2018)
F. peregrinus(f)	Late 10th–late 11th century	St Aldate's Oxford, UK	Poole (2018)

Species	Dates	Location	References
F. peregrinus[f]	1160–1180 CE	Faccombe Netherton, UK	Poole (2018)
F. peregrinus[f]	11th–13th century	Ilchester, UK	Poole (2018)
F. peregrinus[f]	14th century	Beverley, UK	Poole (2018)
F. peregrinus[f]	13th century	Winchester Palace, Southwark, UK	Poole (2018)
F. peregrinus[f]	13th century	Ramsbury, UK	Poole (2018)
F. peregrinus[f]	1350–1510 CE	Southampton, UK	Poole (2018)
F. peregrinus[f]	1500 CE	Bayard's Castle. UK	Poole (2018)
F. peregrinus[f]	12th–15th century	Loughor Castle, UK	Poole (2018)
F. peregrinus[f]	12th–15th century	Castle Rising, UK	Poole (2018)
F. peregrinus[f]	7th–8th century	Brandon, United Kingdom	Crabtree (1996)
F. peregrinus[f]	c. 1500 CE	London, UK	Bramwell (1975)
F. peregrinus[f]	Early 7th century CE	Rickeby, Sweden	Vretemark (2013, 2018)
F. peregrinus[f]	7th century CE	Landshammar, Sweden	Vretemark (2018)
F. peregrinus[f]	7th century CE	Algö, Sweden	Vretemark (2018)
F. peregrinus[f]	7th century CE	Sunnerby, Sweden	Vretemark (2018)
F. peregrinus[f]	7th–8th century CE	Almvägen, Sweden	Vretemark (2018)
F. peregrinus[f]	9th–10th century CE	Broby, Sweden	Vretemark (2018)
F. peregrinus[f]	10th century	Trelleborg, Denmark	Gotfredsen (2014)
F. peregrinus[f]	8th–9th century	Mikulčice, Czech Republic	Mlíkovský (2003)
F. peregrinus[f]	10th–12th century	Budeč, Czech Republic	Peške (1985)
F. peregrinus[f]	Middle Ages	King's Lynn, UK	Bramwell (1977)
F. peregrinus[f]	9th–15th century	Lincoln, UK	O'Connor (1982)
F. peregrinus[f]	9th century (falconry)	Mikulčice Castle, Czech Republic	Goldberg (2020)
F. peregrinus[f]	13th–15th century CE	Tver, Russia	Zinoviev (2021)
F. peregrinus[f]	12th century CE	Novgorod, Russia	Maltby and Hamilton-Dyer (1995)
F. peregrinus[f]	650–750 CE	Salme, Estonia	Peets et al. (2021)
F. peregrinus[f]	14th–16th c. CE	Vilnius Lower Castle, Lithuania	Ehrlich et al. (2023)
F. peregrinus[f]	11th–12th c. CE (falconry)	Valkenburgs, The Netherlands	Prummel (2018)
F. peregrinus[f]	8th c. CE (falconry)	Wellin, Belgium	Prummel (2018)
F. peregrinus[r]	11th–12th c. CE	Zagora, Thessaly, Greece	Koukoules (1932)
F. peregrinus[r]	11th–12th c. CE	Stenimachos, Macedonia, Greece	Koukoules (1932)
Falco sp.[f]	Late 12th–14th century	Brighton Hill South, UK	Poole (2018)
Falco sp.[r]	8th century (trade)	Mercia (Midlands), UK	Dobney and Jaques (2002)
Falco sp.[f]	7th century CE	Staufen, Germany	Schmölcke (2018)
Falco sp.[f]	8th–9th century CE	Hedehusum, Germany	Schmölcke (2018)
Falco sp.[f]	650–750 CE	Salme, Estonia	Peets et al. (2021)
Falco sp.[f]	14th–16th c. CE	Vilnius Lower Castle, Lithuania	Ehrlich et al. (2023)

References

Abeele B (1990) La Fauconnerie dans le lettres française du XII au XIV siècle. Leuven University Press, Leuven, Belgium

Ægisson S (2015) Icelandic trade with Gyrfalcons from Medieval times to the modern era. Published privately Siglufjörður, Iceland

Akasoy A, Georges S (eds) (2005) Das Falken- und Hundebuch des Kalifen al-Mutawakkil. Akademie Verlag, Berlin

Albarella U, Davis SJM (1994) Medieval & post-medieval mammal & bird bones from Launceston Castle, Cornwall; 1961–1982 excavations, HBMC Ancient Monuments Laboratory report, 18/94, London

Allen M (1980) Falconry in Arabia. Orbis, London

Allsen TT (2006) The royal hunt in Eurasian history. University of Pennsylvania Press, Philadelphia

Amari M (1880) Biblioteca Arabo-Sicula. Loescher, Turin/Rome

Anchukaitis KJ, Smerdon JE (2022) Progress and uncertainties in global and hemispheric temperature reconstructions of the Common Era. Quat Sci Rev 286:107537. https://doi.org/10.1016/j.quascirev.2022.107537

Arabadzhiev I (1962) Birds of prey in Bulgaria. Science and Art, Sofia

Arcussia (d') C (1598) La fauconnerie de Charles d'Arcussia, seigneur d'Esparron de Paillieres & de Courmes, Gentilhomme Provençal, divisée en tros livres. Jean Tholosan, Aix

Askeyev IV, Galimova OV, Askeyev OV (2013) Domestic and wild birds from Medieval archaeological sites of Middle Volga region. In: Askeyev IV, Ivanov DV (eds) The dynamics of modern ecosystems in the Holocene. Otechestvo Publishing House, Kazan, pp 72–77

Attiya HM (1999) Knowledge of Arabic in the Crusader States in the twelfth and thirteenth centuries. J Mediev Hist 25:203–213. https://doi.org/10.1016/S1304-4184(98)00022-0

Audoin-Rouzeau F (1986) Ossements animaux du Moyen-Age au monastère de La Charité-sur-Loire. Publ de la Sorbonne, Paris

Battaglia S (1977) De Falconibus et Girofalcis. Sellerio, Palermo

Baumgart W (1980) Der Sakerfalke. Ziemsen, Wittenberg (Lutherstadt)

Becker C (2005) Birds, mice, and slaughter refuse from an Islamic Mosque in Syria—a puzzling mixture at a peculiar location. Doc Archaeobiol 3:271–280

Belon du Mans P (1555) L'histoire de la nature des oyseaux, auec leurs descriptions, & naïfs portraicts retirez du naturel: escrite en sept liures. Guillaume Cauellat, Paris

Bertolucci Pizzorusso V (1975) Marco Polo, Il Milione. Adelphi, Milano

Bleile F (2018) Falconry among the Slavs of the Elbe? In: Gersmann K-H, Grimm O (eds) Raptor and human—falconry and bird symbolism throughout the millennia on a global scale—vol 1/3. Wachholtz, Kiel, pp 1303–1370

Bochenski ZM, Tomek T, Wertz K, Wojenka M (2018) Falconry in Poland from a zooarchaeological perspective. In: Gersmann K-H, Grimm O (eds) Raptor and human—falconry and bird symbolism throughout the millennia on a global scale—vol 1/3. Wachholtz, Kiel, pp 1399–1410

Boev Z (2016) New animal remains from Pliska, the medieval capital (10 c. AD) of Bulgaria, (Shumen Region, NE Bulgaria). ZooNotes 87:1–3

Borreggine M, Latychev K, Coulson S, Powell EM, Mitrovica JX, Milne GA, Alley RB (2023) Sea-level rise in Southwest Greenland as a contributor to Viking abandonment. PNAS 120:e2209615120. https://doi.org/10.1073/pnas.2209615120

Böse H (1973) Thomas of Cantimpré, Liber de natura rerum. De Gruyter, Berlin

Bramwell D (1975) Bird remains from medieval London. London Naturalist 54:15–20

Bramwell D (1977) Bird bone. In: Clarke H, Carter C (eds) Excavations in King's Lynn 1963–1970. Society for Medieval Archaeology, Monograph 7, pp 399–403

Bresc H (1980) La chasse en Sicile (XIIe – XVe siècles). In: Actes du Colloque de Nice: La chasse au Moyen- Age. Publ Fac Lettres Sc Hum Nice, Les Belles Lettres, Paris, pp 201–217

Buquet T (2021) The gyrfalcon in the Middle Ages, an exotic bird of prey (Western Europe and Near East). In: Burnett C, van den Abeele B (eds) Falconry in the Mediterranean context during the pre-modern era. Bibliotheca Cynegetica, Broz, pp 79–98
Carr M (2015) Crossing boundaries in the Mediterranean: papal trade licences from the Registra supplicationum of Pope Clement VI (1342–52). J Mediev Hist 41:107–129. https://doi.org/10.1080/03044181.2014.986869
Carus-Wilson E (1962) The Medieval trade of the ports of the Wash. Mediev Archaeol 6:182–201. https://doi.org/10.1080/00766097.1962.11735666
Coy J (2009) Late Saxon and Medieval Animal Bone from the Western Suburbs. In: Serjeantson D, Rees H (eds) Food, Craft and Status in Medieval Winchester: the plant and animal remains from the suburbs and city defences. Winchester Museums, Winchester
Crabtree PJ (1996) Production and consumption in an early complex society: animal use in Middle Saxon East Anglia. World Archaeol 28:58–79. https://doi.org/10.1080/00438243.1996.9980331
Cummins J (1988) The Hound and the Hawk. Weidenfeld and Nicolson, London
Dalby D (1965) Lexicon of the medieval German hunt. De Gruyter and Co., Berlin
Dobney K, Jaques D (2002) Avian signatures for identity and status in Anglo-Saxon England. Acta Zool Cracov 45:7–21
Duceppe-Lamarre F (2000) Une génération de gestion animale au début du XIV siècle: la compatibilité du territorie d'élevage et de chassed'Hesdin (Pas-de-Calais, France). Athropozoologica 31:125–135
Ehrlich F, Piličiauskienė G, Blaževičius P (2023) Bird remains from Vilnius Lower Castle, Lithuania (13th–19th centuries) reveal changes in social status and unusual bird pathologies. Int J Oestoarchaeol 33:731–741. https://doi.org/10.1002/oa.3193
Emerton E, Noble TFX (2000) The letters of Saint Boniface. Columbia University Press, New York
Evans D (1992) A history of nature conservation in Britain. Routledge, London
Fenech N (2010) A complete guide to the birds of Malta. Midsea, Malta
Fiorini S (2001) The Maltese Falcon, falconry in Malta 1239–1500. Treasures of Malta Vol VII No 2. Fondazzjoni Patrimonju Malti, Malta
Fonteijne H (1993) Het Smelleken (*Falco columbarius*). Fitis 29:122–125
Forcada M (2011) Calendar of Córdoba. In: Fleet K, Krämer G, Matringe D, Nawas J, Rowson E (eds) Encyclopaedia of Islam, THREE. https://doi.org/10.1163/1573-3912_ei3_COM_24375
Fradejas Rueda JM (2008) sha<u>dh</u>āniqat al-balansiyya or sha<u>dh</u>āniqat al-baḥriyya: on the Arabic text and the Latin translations of the Calendar of Cordova. In: Goyens M, de Leemans P, Smets A (eds) Science translated: Latin and vernacular translations of scientific treatises in Medieval Europe. Leuven University Press, Leuven, pp 59–72
Fradejas Rueda JM (2009) La versión catalana de la Epistola aquilae symachi et theodotionis ad ptolomeum regem aegypti y su modelo latino. RLLCGV 14:89–102
Frajedas Lebrero J, Rueda F (2016) Pero López de Ayala, Libro de la caza de las aves. Castalia, Barcelona
Friedland K (1984) The Hanseatic League and Hanse Towns in the early penetration of the North. Arctic 37:539–543
Fritz B (1993) "Falkbrevet" i Barcelona. Riksarkivet, Månadens Dokument, mars 1993. Unpublished report
Gál E (2008a) Vogelfunde aus der Festung von Bajcsavár. In: Kramer D (ed) Auf Sand gebaut Weitschawar/Bajcsa Vár eine steirische Festung in Ungarn. Historischen Landeskomission für Steiermark, Graz, pp 120–124
Gál E (2008b) New data to the bird bone artefacts from Hungary and Romania. In: Luik H, Choyke AM, Batey CE, Lõugas L (eds) From Hooves to Horns, from Mollusc to Mammoth. Manufacture and use of bone artefacts from prehistoric times to the present. In: Proceedings of the 4th meeting of the ICAZ worked bone research group. Muinasaja teadus 15, pp 325–338
Gál E (2015) "Fine feathers make fine birds": the exploitation of wild birds in Medieval Hungary. Antaeus 33:345–368

Gál E, Daróci-Szabó L, Daróczi-Szabó M (2020) Bird bone remains from two medieval settlements in Debrecen (Eastern Hungary). Acta Archaeol Acad Sci Hung 71:199–218. https://doi.org/10.1556/072.2020.00008

Galimova DN, Askeyev IV, Askeyev OV (2014) Bird remains from 5th-17th century AD archaeological sites in the Middle Volga region of Russia. Int J Osteoarchaeol 24:347–357. https://doi.org/10.1002/oa.2385

George W, Yapp B (1991) The naming of the beasts. Duckworth, London

Girininkas A, Daugnora L (2018) Premodern hunting with birds of prey in the historical Lithuanian lands: entertainment, politics or economy necessity? In: Gersmann K-H, Grimm O (eds) Raptor and human—falconry and bird symbolism throughout the millennia on a global scale—vol 1/3. Wachholtz, Kiel, pp 1215–1227

Goldberg EJ (2020) In the manner of Franks. University of Pennsylvania Press, Philadelphia

Gorobets L, Kovalchuk O (2017) Birds in the medieval culture and economy of the East Slavs in the 10–13th centuries AD. Environ Archaeol 22:147–165. https://doi.org/10.1080/14614103.2016.1141088

Gotfredsen AB (2014) Birds in Subsistence and Culture at Viking Age Sites in Denmark. Int J Osteoarchaeol 24:365–377. https://doi.org/10.1002/oa.2367

Govaerts S (2024) Biodiversity in the Late Middle Ages: wild birds in the fourteenth-century county of Holland. Environ Hist 30. https://doi.org/10.3197/096734022X16627150608122

Hamilton-Dyer S (2002) The bird resources of Medieval Novgorod, Russia. Acta Zool Cracov 45(special issue):99–107

Harrison R (2009) The Gásir area A archaeofauna: an update of the results from the faunal analysis of the High Medieval trading site in Eyjafjörður, N Iceland. NORSEC Zooarchaeology Laboratory Report No. 44

Harrison R, Roberts HM, Adderley WP (2008) Gásir in Eyjafjörður: international exchange and local economy in Medieval Iceland. J North Atlantic 1:99–119. https://doi.org/10.3721/1935-1933-1.1.99

Hartman S, Ogilvie AEJ, Ingimundarson JH, Dugmore AJ, Hambrecht G, McGovern TH (2017) Medieval Iceland, Greenland, and the new human condition: a case study in integrated environmental humanities. Glob Planet Chang 156:123–139. https://doi.org/10.1016/j.gloplacha.2017.04.007

Haskins CH (1971) The renaissance of the twelfth century. Harvard University Press, Cambridge

Heckmann M-L, Heckmann D (2017) Falken im Ordensland und Herzogtum Preußen—ein ökologisch und ökonomisch knappes Gut. Preußenland 8:7–40

Huillard-Bréholles JL (1852–1861) Historia diplomatica Friderici Secundi, sive constitutiones, privilegia, mandata, instrumenta quae supersunt istius Imperatoris et filiorum ejus. Accedunt epistolae Paparum et documenta varia, 7 vols in 12 parts, Paris

Jánossy D (1985) Wildvogelreste aus archäologischen Grabungen in Ungarn (Neolithicum bis Mittelalter). Fragmenta Mineralogica et Palaeontologica 12:67–103

Jobling JA (2011) The Helm disctionary of scientific bird names. C Helm, London

Koukoules P (1932) Κυνηγετικά εκ της εποχής των Κομνηνών και των Παλαιολόγων [Hunting in the eras of the Comneni and the Palaiologoi]. Yearbook Soc Byzantine Stud 1932:3–33

Kumerloeve H (1963) Le Lanier *Falco biarmicus* Temminck nichait-il en France et en Espagne il y a quelques siècles? Alauda 31:1–4

Lambert J (2008) Medieval translations and translation studies: some preliminary considerations. In: Goyens M, de Leemans P, Smets A (eds) Science translated: Latin and Vernacular translations of scientific treatises in Medieval Europe. Leuven University Press, Leuven, pp 1–10

Landro T (2019) Periodization and temporal boundaries—defining the Middle Ages. In: Orban F, Strand Larsen E (eds) Discussing borders, escaping traps. Waxmann, Münster, pp 17–34

Landsborough Thomson A (1976) The name of Eleonora's Falcon. Bull BOU 96:112

Leonardi G (2015) The Lanner falcon. GLE, Catania

Leonardi G, Sutton LJ (2020) *Falco lanarius*: the contorted history of a poetical archetype and a taxonomic puzzle. Archaeol Anthropol Sci 12:22. https://doi.org/10.1007/s12520-019-00995-1

Lie RO (2018) Falconry, falcon-catching and the role of birds of prey in trade and alliance gifts in Norway (800–1800 AD) with an emphasis on Norwegian and later foreign participants in falcon catching. In: Gersmann K-H, Grimm O (eds) Raptor and human—falconry and bird symbolism throughout the millennia on a global scale—vol 1/2. Wachholtz, Kiel, pp 727–786

Lindner K (1962) Von Falken, Hunden und Pferden: deutsche Albertus-Magnus- Übersetzungen zu der ersten Hälfte des 15. Jahrhunderts. de Gruyter, Berlin

Lupino G (2022) Carte de Logu dell'Arborea. Il Maestrale, Nuoro

Luterbacher J, Werner JP, Smerdon JE, Fernández-Donado L, González-Rouco J, Barriopedro D et al (2016) European summer temperatures since Roman times. Environ Res Lett 11:024001. https://doi.org/10.1088/1748-9326/11/2/024001

Makowiecki D, Makowiecka M (2013) The character of animal exploitation and the environment at the Polish/Prussian frontier in the Medieval Period: a case study. Archaeologia Baltica 20:91–116

Maltby M, Hamilton-Dyers S (1995) Animal bones from excavations in Novgorod the Great and its vicinities, Novgorod and Novgorod Region. Hist Archaeol 9:129–157

Marruncheddu S (2008) La traduction française du Moamin dans ses rapports avec la version latine de Théodore d'Antioche. In: Goyens M, de Leemans P, Smets A (eds) Science translated: Latin and vernacular translations of scientific treatises in Medieval Europe. Leuven University Press, Leuven, pp 297–310

Martinez Enamorado V (2011) Falcons and falconry in Al-Andalus. Studia Orientalia 111:159–183

Masseti M (2009) In the gardens of Norman Palermo, Sicily (twelfth century A.D.). Anthopozoologica 44:7–34

Mayaud N (1963) Observations et documents nouveaux. *Falco biarmicus* Temminck. Faucon Lanier. Alauda 31:39–40

McGovern TH, Bigelow GF, Amorosi T, Woollett J, Perdikaris S (1993) The zooarchaeology of Ø 17a. In: Vebæk CL (ed) Narsaq—a Norse landnáma farm. Meddelelser om Grønland Man & Society 18. Danish Polar Center, Copenhagen, pp 58–74

McGovern TH, Amorosi T, Perdikaris S, Woollett J (1996) Vertebrate zooarcheaeology of Sandnes V51: economic change at a chieftain's farm in west Greenland. Arct Anthropol 33:94–121

Mlíkovský J (2003) Die Vögel aus der frühmittelalterlichen Burg Mikulčice, Mähren. In: Polácek L (ed) Studien zum Burgwall von Mikulčice 5, pp 215–338

Mlíkovský J (2004) Zvirata ajejich rolena ranc stredovékém hrade Stará Boleslav (Stiedni Cechy). In: Bohácová I (ed) Stará Boleslav Piemyslovsky hrad v ranémstiedoveku. Medievalia Archaeologica 5, pp 347–365

Moreno-García M, Pimenta C (2010) Beyond chicken: avian biodiversity in a Portuguese late medieval urban site. Groningen Archaeol Stud 12:261–275

Moseley GE, Rosvold J, Gotfredsen AB, Hajdas I, Gilg O, Gregersen KM, Spötl C, Edwards RL (2019) First pre-modern record of the Gyrfalcon (*Falco rusticolus*) in Northeast Greenland. Polar Res 38:3539. https://doi.org/10.33265/polar.v38.3539

Mulkeen S, O'Connor TP (1997) Raptors in towns: towards an ecological model. Int J Oesteoachaeol 7:440–449. https://doi.org/10.1002/(SICI)1099-1212(199707/08)7:43.0.CO;2-3

Müller H-H (1993) Falconry in central Europe in the Middle Ages. In: Exploitation des animaux sauvages a travers le temps. XIII Rencontres Int d'Archeologie et d'Historie d'Antibes. Editions APDCA, Juan-les-Pines, pp 431–437

Myers S (2022) The bird name book. Princeton University Press, Princeton/Oxford

Negro JJ, Prenda J, Ferrero JJ, Rodríguez A, Reig-Ferrer A (2020) A timeline for the urbanization of wild birds: the case of the lesser kestrel. Quat Sci Rev 249:106638. https://doi.org/10.1016/j.quascirev.2020.106638

O'Connor TP (1982) Animal bones from Flaxengate, Lincoln c 870–1500. Council for British Archaeology, London

Oggins RS (1989) Falconry and medieval social status. Medievalia 12:43–55

Oggins RS (1993) Falconry and medieval views of nature. In: Salisbury JE (ed) The medieval world of nature. Garland Publications, New York/London, pp 47–60

Oggins RS (2004) The Kings and their hawks. Yale University Press, New Haven/London

Peets J, Allmäe R, Maldre L, Saage R, Tomek T, Lõugas L (2021) Research results of the Salme ship burials in 2011–2012. Archaeol Fieldwork Estonia 2012:43–60
Peške L (1985) Domáci a lovná zvitata podle nálezú na slovanskrch lokalitách v Cechách.—Sbornik Národniho Musea v Praze. Rada Historická 39:209–216
Peške L (1993) Nálezy kosti ptákú z Cech a Moravy z doby po poslednim zaledněni. Zprávy Ceské Spolecnosti Ornitologické 36:53–58
Petersen A (1998) Icelandic birds. Vaka-Helgafell, Reykjavík
Pickering OS, O'Mara VM (1999) Manuscripts in Lambeth Palace Library, including those formerly in Sion College Library. Boydell & Brewer, Cambridge
Poole K (2018) Zooarchaeological evidence for falconry in England, up to AD 1500. In: Gersmann K-H, Grimm O (eds) Raptor and human—falconry and bird symbolism throughout the millennia on a global scale—vol 1/3. Wachholtz, Kiel, pp 1027–1053
Popescu SM, Tigae C, Dobrițescu A, Ştefănescu DM (2024) Exploring the climatic niche evolution of the genus *Falco* (*Aves*: *Falconidae*) in Europe. Biology (Basel) 13:113. https://doi.org/10.3390/biology13020113
Prummel W (2013) Falconry in continental settlements as reflected by animal bones from the 6th to 12th centuries AD. In: Grimm O, Schmölcke U (eds) Hunting in northern Europe until 1500 AD. Schr Arch Landesmus Ergbd 7, pp 357–378
Prummel W (2018) The archaeological-archaeozoological identification of falconry – methodological remarks and some Dutch examples. In: Gersmann K-H, Grimm O (eds) Raptor and human—falconry and bird symbolism throughout the millennia on a global scale—vol 1/2. Wachholtz, Kiel, pp 467–478
Prummel W, Heinrich D (2005) Archaeological evidence of former occurrence and changes in fishes, amphibians, birds, mammals and molluscs in the Wadden Sea area. Helgol Mar Res 59:55–70. https://doi.org/10.1007/s10152-004-0207-1
Prummel W, Küchelmann HC (2022) The use of animals in settlements on the Dutch and German Wadden Sea coast, 600 BC – AD 1500. In: Nicolay J, Schepers M (eds) Embracing the salt marsh: foraging, farming and food preparation in the Dutch-German coastal area up to AD 1600m. Jaarverslagen van de Vereniging voor Terpenonderzoek 102/Groningen Archaeological Studies 42. Groningen, pp 109–127
Schlegel H (1844) Revue critique des oiseaux d'Europe. Roret, Paris
Schmölcke U (2018) Central Europe burials with birds of prey from the middle of the 1st millennium AD—a short survey of the early history of archaeozoology in connection with these burials. In: Gersmann K-H, Grimm O (eds) Raptor and human—falconry and bird symbolism throughout the millennia on a global scale—vol 1/2. Wachholtz, Kiel, pp 495–505
Schmölcke U (2022) What about exotic species? Significance of remains of strange and alien animals in the baltic sea region, focusing on the period from the Viking Age to High Medieval Times (800–1300 CE). Heritage 5:3864–3880. https://doi.org/10.3390/heritage5040199
Serdyuk NV, Syromyatnikova EV, Zelenkov NV, Abdykanova A, Kyzy SA, Shnaider SV (2023) Holocene vertebrate fauna in Fergana Valley, Kyrgyzstan, based on fossils from the Obishir-5 rock shelter. Geobios 76:37–51. https://doi.org/10.1016/j.geobios.2023.01.002
Serjeantson D (2009) Birds—Cambridge manuals in archaeology. Cambridge University Press, Cambridge
Shaymuratova D, Askayev I, Askayev O (2019) The studies of archaeological bird remains from Medieval Staraya Ladoga. MASF 7:93–114
Shaymuratova D, Askeyev A, Askeyev O, Askeyev I (2023) Birds of prey from 4th to 18th centuries AD of the Volga River basin of Russia. Int J Osteoarchaeol 33:742–752. https://doi.org/10.1002/oa.3168
Shehada HA (2012) Mamluks and animals. Brill, Leiden
Sidiropoulos K, Polymeni R, Legakis A (2017) The evolution of Greek fauna since classical times. Hist Rev/La Revue Historique 13:127–146. https://doi.org/10.12681/hr.11559

Smets A (2002) La réception en langue vulgaire du "De falconibus" d'Albert le Grand. In: Donavin D, Poster C, Utz R (eds) Medieval forms of argument: disputation and debate (Disputatio 5). Wipf and Stock Publishers, Eugene, pp 189–199
Stadler H (1916–1920) Albertus Magnus, *De animalibus*. Aschendorf, Münster
Standley E (2008) Ladies hunting: a late medieval decorated mirror case from Shapwick, Somerset. Antiquar Jour 88:198–206. https://doi.org/10.1017/S0003581500001396
Sten S, Vretemark M (1988) Storgravsprojektet, osteologiska analyser av yngre ja Èrnalderns benrika brandgravar. ForvaÈnnen 83:145–156
Stuhmiller J (2015) Hunting, hawking, fowling, and fishing. In: Classen A (ed) Handbook of Medieval culture, vol 2. De Gruyter, Berlin, pp 697–721
Tangl M (1916) Die Briefe Des Heiligen Bonifatius Und Lullus.Berlin, pp 229–231
Teegen W-R (2018) The skeletons of a peregrine and a sparrowhawk and the spatial distribution of birds of prey in the Slavonic fortification of Starigard/Oldenburg (Schleswig-Holstein, northern Germany, 7th–13th centuries). In: Gersmann K-H, Grimm O (eds) Raptor and human—falconry and bird symbolism throughout the millennia on a global scale—vol 1/3. Wachholtz, Kiel, pp 1371–1398
Tkacz MW (2009) Albert the Great and the Aristotelian reform of the Platonic method of division. Thomist 73:399–435. https://doi.org/10.1353/tho.2009.0013
Trombetti Budriesi AL (2000) De arte venandi cum avibus. L'arte di cacciare con gli uccelli. Edizione critica e traduzione italiana del ms. lat. 717 della Biblioteca Universitaria di Bologna collazionato con il ms. Pal. Lat. 1071 della Biblioteca Apostolica Vaticana. Laterza, Roma-Bari
Tyrberg T (2002) The archaeological record of domesticated and tamed birds in Sweden. Acta Zool Cracov 45:215–231
Vaughan R (1982) The Arctic in the Middle Ages. J Mediev Hist 8:313–342
Viré F (1965) La Volerie dans l'Espagne du Xe siècle à travers le calendrier de Cordue. Arabica 12:306–314
Viré F (1966) Le traité de l'art de volerie (Kitāb al-Bayzara) rédigé vers 385/995 par le Grand-Fauconnier du calife fāṭimide al-'Azīz bi-llāh: IV. Arabica 13:39–76
Viré F and Möller D (2002) al-Ghitrīf, Traité des oiseaux de vol (Kitāb Ḍawari aṭ-ṭayr). Le plus ancien traité de fauconnerie arabe. Jacques Laget, Nogent-le-Roi, p 84 and note 127
von Strassburg G (1974) Tristan. Penguin, London
Vretemark M (1997) Från ben till boskap. Kosthåll och djurhållning med utgångspunkt i medeltida benmaterial från Skara. Vol 1. Skrifter från Skaraborgs Länsmuseum, 25. Skaraborgs länsmuseum, Skara
Vretemark M (2013) The Vendel Period royal follower's grave at Swedish Rickeby as starting point for reflections about falconry in Northern Europe. In: Grimm O, Schmölcke U (eds) Hunting in northern Europe until 1500 AD: old traditions and regional developments, continental sources and continental influences. Wachholtz, Neumünster, pp 379–386
Vretemark M (2018) Birds of Prey as evidence for falconry in Swedish burials and settlements (550-1500 AD). In: Gersmann K-H, Grimm O (eds) Raptor and human—falconry and bird symbolism throughout the millennia on a global scale—vol 1/2. Wachholtz, Kiel, pp 827–839
Walker SJ, Hufthammer AK, Meijer HJM (2019) Birds in Medieval Norway. Open Quat 5:5. https://doi.org/10.5334/oq.58
Yapp WB (1979) The birds of English medieval manuscript. J Mediev Hist 5:318–348
Yapp WB (1982) Birds in captivity in the Middle Ages. Arch Nat Hist 10:479–500
Yapp WB (1983) The illustrations of birds in the Vatican manuscript of De arte venandi cum avibus of Frederick II. Ann Sci 40:597–634. https://doi.org/10.1080/00033798300200401
Yapp WB (1987) Animals in Medieval Art: the Bayeux Tapestry as an example. J Mediev Hist 13:15–73. https://doi.org/10.1016/0304-4181(87)90042-X
Zambon F (2018) Bestiari tardoantichi e medievali. Bompiani, Florence
Zinoviev AV (2021) Zooarchaeology of Tver Kremlin (12th–18th centuries, Tver, Russia). Int J Osteoarchaeol 31:568–582. https://doi.org/10.1002/oa.2972

Chapter 5
Early Modern Period

5.1 The Birth of Ornithology

This 300-year period marks the beginning of ornithological knowledge and, to some extent, the study of falcons. Influenced by Albert the Great's Aristotelian approach from 200 years earlier, a new way of looking at nature emerged, offering a fresh perspective on the animal and plant world, even if it remained somewhat verbose. Additionally, the study of animals was stimulated by the publication of the zoological writings of Aristotle and Pliny (Perfetti 2000). William Turner (1508–1568) from Northumberland wrote "*Avium praecipuarum quarum apud Plinium et Aristotelem mentio*" (1544), likely the first scientific book on the natural history of birds rather than their practical uses. He arranged birds based on the names and observations of Aristotle and Pliny, although other authors preferred to follow falconry treatises for falcons. Albertus Magnus's detailed classification of species was so convincing and rich in experience that its influence extended well into the early modern era of ornithology (see Sects. 5.2 and 5.3).

In the sixteenth century, ornithology matured into a respected science in Europe. Concurrently, the quality of paintings featured in manuals significantly improved with the use of life models, both deceased or living (Olson and Mazzitelli 2007). For example, Konrad Gessner's "*Historiae animalium*" was the first well-illustrated animal book, though some illustrations still leaned toward the fantastical. Ulisse Aldrovandi (1522–1605) titled one of his most famous bird treaties "*Ornithologiae*," deriving its name from the New Latin word "*ornithologia*," which originated from the Ancient Greek "*ornithológos*" (ὀρνιθολόγος; Fig. 5.1). Although highly regarded in his time, his later contributions to natural sciences faced significant criticism (Egerton 2003). Nevertheless, the illustrations from his bird volumes (e.g., Fig. 5.3) remained influential a century after their publication (Ellenius 1997).

This theoretical approach to understanding the natural world has not solely focused on enhancing observation techniques. Scientists frequently communicated

G. Leonardi, *The History of Western Palearctic Falcons*,
https://doi.org/10.1007/978-3-031-86528-2_5

Fig. 5.1 The frontispiece of Ulisse Aldrovandi's "*Ornithologiae*". (Aldrovandi 1599; Source: BHL https://www.biodiversitylibrary.org)

with one another, leading to lively discussions about important discoveries and encouraging critical thinking about expected outcomes (Hatch 2000). Unlike other animal groups, the wealth of knowledge gained from falconry provided a highly credible foundation for naturalists of the era. During the Renaissance, falconry reached its peak popularity in both the West and the Islamic world (Arbel 2009). Bone remains of large falcons, such as the peregrine falcon, have been found in castles, while smaller falcons, such as common kestrels and hobbies, are also found in towns (Zeiler 2010; see Appendix). However, the presence of captive falcons has greatly complicated the accurate assessment of native species within a given territory. For instance, in his natural history work "*Scotia Illustrata*" (1684), Robert Sibbald includes the saker falcon among the native species (Raye 2018). Even during the early modern period, monarchs such as the German emperors, the kings of England, France, Spain, and the Ottoman sultans relished their time with falcons and hawks (Arbel 2009). Consequently, the extensive falcon trade has left behind many administrative and diplomatic documents, which should be interpreted carefully.

In the eighteenth century, Ray and Willoughby's encyclopedia sparked a surge in systematic studies, which continued consistently into the nineteenth century (see Sect. 5.3; Ray 1678). A major contribution came from Carl Linnaeus (1707–1778) with his publication "*Systema Naturae*," which emphasized genera and introduced the use of binomial nomenclature (see Sect. 5.4; Linnaeus 1758). Conversely, George Louis Leclerc, later Comte de Buffon, in his book "*Histoire naturelle*," proposed that animal species do not persist the same over time but are influenced by factors such as climate or diet (see Sect. 5.4; Buffon 1785). His book, written in accessible language rather than Latin, achieved widespread popularity among a broad audience (Birkhead and Charmantier 2009). Both texts significantly impacted efforts to collect specimens for classification and contributed to the exponential growth in the number of taxa (see Chap. 6; Gassó Miracle 2008).

5.2 The Sixteenth Century

Three literary sources from this century will provide the foundation of knowledge for future historical periods, extending to modern times. These works were primarily written in Latin, the language used by scientists, with the exception of Belon (1555), who wrote in his native French (Birkhead and Charmantier 2009). Konrad Gessner (1515–1565) authored "*Historiae animalium liber III*," focusing on birds and using a classification based on falconry treatises rather than Aristotle's (Gessner 1554). He also faithfully followed Albertus Magnus's classification as reported in "*De falconibus*" (see Sect. 4.4.2). Despite this, Gessner's work remains one of the foundational sources for the critical review later undertaken in Ray and Willoughby's encyclopedia (see Sect. 5.3).

Pierre Belon (1517–1564), a pioneer of modern ichthyology, authored "*Histoire de la Nature des Oyseaux*" (Belon 1555), a seminal work on birds. He continued the falconry tradition by using names derived from it to distinguish between male and female birds (such as "*sacre-sacrette*," "*lanner-lannerette*," etc.). Belon classified birds based on their practical skill: "*faucon pèlerin*," "*faucon de Tartarie*" (= Asia), "*faucon de Barbarie*," "*faucon gerfault*," "*faucon sacre*," "*faucon lanier*," "*faucon tunicien*," "*emerillon*," and "*hobreau*" (Belon 1555). He also noted that gyrfalcons, which come from Russia, are caught during migration in Germany along with peregrine falcons and then transported to France (see also Sect. 4.3.2). Belon further observed that saker falcons, originating from Tartary and Russia, are caught on islands in the Aegean Sea (such as Rhodes, Karpathos, Crete, and Naxos) and in Cyprus (Belon 1555). Today, migration patterns indicate that few saker falcons visiting these islands likely originate from the farthest eastern regions of Europe (European Russia and western Kazakhstan) and the farthest western regions of Asia (Eastern Turkey; Prommer et al. 2012). For the common kestrel, Belon drew on information from Aristotle and Pliny (cf. Sect. 3.6.1) and also mentioned the Italian name from that time, "*foutivento*," which refers to hunting rodents using soaring flight (Belon 1555; Fig. 5.2).

In his book "*Ornithologiae*," Aldrovandi collects and combines information from Greek and Latin authors like Aristotle and Pliny, as well as treatises on falconry by Frederick II and Albertus Magnus (Aldrovandi 1599). He also refers to a translated copy of Belon's book. Due to its comprehensive nature, "*Ornithologiae*" became a fundamental text, particularly regarding falcons, even in the following century (see Sect. 5.3). Much emphasis is placed on highly valued species, such as the gyrfalcons, which were acquired through trade with the North (see Sect. 4.2.3). In his 1561 treatise "*La venerie*," Jacques du Fouilloux reports that Ivan the Terrible, the Tsar of Russia, employed gyrfalcons for hunting ravens. In the late sixteenth century, Jacques Auguste de Thou stated that the gyrfalcon lives in Scandinavia and the mythical island of Thule (possibly Iceland). Aldrovandi specifically mentioned the Norwegian gyrfalcons once owned by Alfonso II d'Este, Duke of Ferrara (1533–1597), which served as the inspiration for the woodcut illustration in his "*Ornithologiae*" (Aldrovandi 1599; Fig. 5.3).

In the sixteenth century, falconry literature not only preserved past knowledge but also integrated new field experiences within its texts. In his manual, Charles d'Arcussia (1554–1628), who served as the first consul of Aix-en-Provence, employed field captures to identify falcon species and subspecies (Arcussia 1598). For instance, the "*faucon gentil*" is captured between July and September, while the "*faucon pelerin*" is captured from mid-September to January. After that, "*au commencement de l'année il perd son nom*" (at the beginning of the year, it loses its name). Arcussia suggests that "*...le difference, il n'y en a point d'autre que celle du temps qui leur fait changer de nom, encores qu'ils soient de meme espece*" (the only distinction lies in the passage of time, prompting them to change their name despite belonging to the same species; Arcussia 1598). Regarding a rare and challenging species like the lanner falcon, Arcussia is the first author to accurately identify its actual presence and nesting sites in the Palearctic region. He notes "*L'abondance des laniers viet de Sicile; et font la plus-part leurs aires dans degráds rochers, et par fois aussi au haut de quelque grand arbre. Il en vient aussi de la Poïlle, lesquels se prennent aux montaignes du pays*" (Arcussia 1598). Remarkably, he mentions that most of these birds originate from Sicily, where they nest among rocks or in the branches of tall trees, as well as from the rocky areas of Apulia in mainland Italy (Arcussia 1598).

5.2.1 Falcons Trade

Alongside the efficiently structured and lucrative falcon market originating from Iceland and Scandinavia, historical sources reveal additional routes where this market flourished. In the early sixteenth century, collecting falcons for export was a regular activity on Kythera, an island in Greece across the south-eastern tip of the Peloponnese peninsula (Arbel 2009). As in other cases, the administrative records from the Republic of Venice fail to specify the species of falcons being exported from the island (Arbel 2009). Arbel (2009) identifies these falcons as Eleonora's

Fig. 5.2 The "*crecerelle*" (= common kestrel), depicted in Belon (1555). (Source: BHL https://www.biodiversitylibrary.org)

Fig. 5.3 On the left is the "*falcone peregrino*" (peregrine falcon). On the right is a watercolor of the Norwegian gyrfalcon belonging to Alfonso II d'Este, Duke of Ferrara. (Aldrovandi 1599; De Smet 2018; Source: BHL, https://www.biodiversitylibrary.org)

falcons due to the substantial number exported and the presumed capture time during their breeding season. This species, known for its colonial behavior, could explain the high numbers. Additionally, since they breed late (July–August), it clarifies their arrival in Venice during autumn (Arbel 2009). Ultimately, falcons were predominantly gathered from the small islet of Avgò, as documented by a seventeenth-century traveller (Arbel 2009). Remarkably, these birds have continued to nest there to the present day. Kythera played a role in a larger supply network stretching from Cyprus, through Crete, to the Venetian colonies in the Adriatic Sea, and onward to the Ottoman Empire (Arbel 2009; Coureas 2017). While Eleonora's falcons are not mentioned in falconry history, other large falcons like saker and peregrine falcons are (Coureas 2017).

Falcons continued to be transported along established routes from earlier times, passing through significant hubs such as Lynn but predominantly through Hamburg (see also Fig. 4.4; Holterman 2020). During the Early Modern period, the trade peaked in 1764, with approximately 200 birds exported (Petersen 1998). Prior to this, the falcon trade was a relatively minor, seasonal business (Grassby 1997). For instance, in the summer of 1593, Thomas Emmerson undertook a 6-week journey to Norway to procure gyrfalcons (Grassby 1997). Similarly, the population of merlins imported from Ireland to Bristol declined due to excessive hawking (Grassby 1997).

Valkenswaard, a municipality and a town in the southern Netherlands near the Belgian border, emerged as the center of the falcon trade in the eighteenth century (van de Wall 2004). While peregrine falcons were captured locally, expeditions were also undertaken to Iceland, Norway, Sweden, and western Russia, extending as far north as Murmansk, to capture gyrfalcons (van de Wall 2004).

The gyrfalcon trade from Scandinavia and Iceland continued until the late seventeenth century and was formally abolished in 1810 (Fig. 5.4; Petersen 1998). By the end of the nineteenth century, Naumann et al. (1899) remarked on the decline of this

Fig. 5.4 Caring for a gyrfalcon in Scandinavia, as depicted in an illustration from the 1555 book "*Historiae de gentibus septentrionalibus*" by Olaus Magnus, the Archbishop of Uppsala. (Source: BHL https://www.biodiversitylibrary.org)

trade: "*In recent times, many falconries great lords have either fallen into disrepair or ceased to exist altogether. The wars in Europe, especially the changes in ownership of these lands, have disrupted the trade, making the hunting of and the species described here increasingly rare.*"

5.3 The Seventeenth Century

Ray and Willoughby's "*The Ornithology of Francis Willoughby*" marked significant advancement in the study of birds by moving away from folklore and focusing primarily on biological observations (Birkhead and Charmantier 2009). However, the section on falcons closely resembles that in Ulisse Aldrovandi's "*Ornithology*." Ray acknowledges, "*we have borrowed from Aldrovandus the descriptions of the several sorts of falcons without omitting any*" (Aldrovandi 1599; Ray 1678). Ray also notes that Willughby's untimely death prevented any opportunity to update knowledge on falcons (Ray 1678). Despite the verbosity of the text, it outlines an initial attempt to categorize falcon species based on both morphological and ecological criteria (Birkhead and Charmantier 2009). Nevertheless, the primary classification criterion remains the falcons' abilities in falconry. Thus, the peregrine falcon is described first, being regarded as the species with the greatest qualities for this sport (Ray 1678). The text cites various peregrine nesting sites based on the authors' knowledge, including Scotland, the Isle of Anglesey in Wales, and the Bologna region in Italy, according to Aldrovandi (Ray 1678). Interestingly, Aldrovandi claimed there were no significant differences in coloration between the species he described and the one known as "*falconarius*" in Germany (Aldrovandi 1599; Ray 1678). For the saker falcon, Ray preferred Frederick II's description, noting that it is smaller than the gyrfalcon but larger than the peregrine falcon (Ray 1678). However, he does not provide details on its distribution and ecology. The name "*gyrfalcon*" was already well understood (see Sect. 4.4.2), and it was believed that there were "*different kinds, distinguished by the colours of their feathers*" (Ray 1678). Ultimately, the classification reflects that of Albertus Magnus from 400 years earlier. Some species, such as the "*gentle*" and "*haggard*" falcons, are simply variations of the peregrine falcon. The barbary falcon from Tunisia is mentioned, though its exact classification remains uncertain (*F. p. pelegrinoides* or *F. b. erlangeri*?).

Detailed information is provided about common small falcons (Ray 1678). The hobby, also known as "*subbuteo*" according to Aldrovandi, is a migratory bird that breeds in England and feeds on larks (Fig. 5.5). The kestrel ("*tinnunculus*" or "*cenchris*") feeds on beetles and rodents and nests in holes, similar to jackdaws. Males are smaller and have ash-colored heads and backs. Even in the case of the merlin ("*aesalon*"), reversed size dimorphism is noted, with the male being smaller and having bluer feathers on the rump near the tail (Ray 1678).

As Charles D'Arcussia did in the previous century, Diogo Fernandes Ferreira in 1616 focused more on studying the behaviors and feeding habits of falcons than on their formal naming (Almaça 1997). He recognized both the small falcons (merlin,

74 De Auibus

DE LITHOFALCO ET DENDROFALCO.

Figura hæc Dendrofalci eſt.

PRAETER ſupradicta falconum genera (inquit Albertus) duo adhuc apud nos reperiuntur, falco lapidarius, (Germanis ein Steinfalck: pro quo nos lithofalci nomen fingimus) & arborarius, (Germanis ein Baumfalck, uel Baumfelckle, poſterius diminutiuum eſt & mari potius conuenit, ut primitiuum fœminæ. quæ differentia in lithofalco etiam & cæteris omnibus accipitrum falconumq; generibus in lingua noſtra obſeruari poteſt. Nos dendrofalcũ dicere uoluimus.) Lapidarius quidem mediæ quantitatis eſt & uigoris inter peregrinum & gibboſum: & inuenitur in alpium rupibus: & eſt eiuſdem nutritionis cum peregrino, Albert.. Celon, (çeſalon forte) uocatur falco lapidum, quòd in præruptis petris pullificet, Idem.

¶ Arborarius uerò medius eſt quantitate & uigore inter gibboſum & mirle (merillum:) & regitur ac nutritur ſimiliter ut mirle, Albert. Dendrofalcus auis egregia & nobilis eſt, ſpecie nõ admodum diſſimilis ſperuerio, nigrior tamen, aliquanto & minor. & quanquam propter paruitatem infirmitatemq; uirium nõ admodum eius uſus eſt in aucupio, eſt tamen auis planè amabilis & ingenio mitis adeò ut emiſſa per ſyluas & agros ad dominum ſuum redeat. Iucundiſsimum quidem illud ſpectaculum eſt quod conflictu ſuo cum monedulis exhibet, Stumpſius. Dendrofalcus, quem ipſe manibus tractaui, à roſtro ad caudæ finem longus erat palmos quatuor, id eſt digitos ſedecim. pedes erãt pallidi, tanquam ex ſubflauo & uiridi colore mixto. dorſum nigrum, ſed margines infimos pennarum capitis & dorſi, præſertim inferioris, ſemicirculi ſubruffi ambiebant. Pennæ alarum nigriores erant: & latus ala

Fig. 5.5 The "*dendrofalco*" (tree falcon), illustrated in Gessner's 1554 work, was identified by Willoughby and later by Ray in 1678 as the Eurasian hobby. (Source: BHL https://www.biodiversitylibrary.org)

Eurasian hobby) and also noted the presence of several subspecies of peregrine falcons in Portugal. Based on the morphological characteristics and capture dates, he likely mentioned both the nominal subspecies, *F. b. brookei* and *F. p. calidus* (Almaça 1997). Interestingly, he identified the barbary falcon (*F. p. pelegrinoides*) breeding on the Cape Verde Islands but confused it with the *brookei* subspecies (Almaça 1997).

5.4 The Eighteenth Century

Similar to the classification of other animal species, the taxonomy of falcons is well-established, tracing back to their categorization in Linnaeus' "*Systema Naturae*" (Linnaeus 1758). Over the past 300 years, knowledge has accumulated rather than significantly changed. When comparing the taxonomy and nomenclature of falcons from the pre-Linnaeus period to Linnaeus' contributions and subsequent modifications, the conservative nature of this field becomes evident (Table 5.1). Linnaeus' work marked the initial definition of the *Falco* genus, including five species of both large and small falcons. Three species, the common kestrel, Eurasian hobby, and gyrfalcon, were described based on direct observation. Others, such as the merlin, already had different names, while the red-footed falcon was distinctly identified for the first time (Table 5.1).

Describing other species is challenging due to confusion over alleged subspecies, size differences between males and females, and the rarity of some species. For instance, the description of "*Falco lanarius*" (lanner falcon) closely resembles that of a gyrfalcon, likely derived from a *juvenile* bird killed in Sweden. Aldrovandi (1599) based classification of the "*lanarius*" according to their geographical origins, a subdivision that Linnaeus maintained, distinguishing between presumed Italian ("*nostrali*") and French ("*forestieri*") birds (Linnaeus 1758; Lorenzi et al. 1768).

In his work, Buffon took a more ecological and ethological approach, focusing on understanding how species interact with their environment (Buffon 1785; Birkhead and Charmantier 2009). Despite this shift, the legacy of falconry remains evident. When describing the gyrfalcon, he outlines the broad migration route from Iceland to Turkey (Buffon 1785; see also Fig. 4.4). He reviews observations and knowledge about the lanner falcon, concluding that no one has truly seen it and that, although once abundant, it is now scarce (Buffon 1785). Interestingly, he notes similarities between the lanner falcon and the saker falcon in both behavior and morphology (Buffon 1785).

In 1771, Marmaduke Tunstall (1743–1790) privately published "*Ornithologia Britannica*" (Tunstall 1771; Boyd and Jessop 1998). In this four-page folio, he defined the peregrine falcon as *Falco peregrinus* and was the first to apply the name *Falco aesalon* to the merlin (Boyd and Jessop 1998). In 1790, John Latham adopted the term "*calidus*," now used to define the northern subspecies of peregrine falcon, *F. peregrinus calidus* (Latham 1790).

By the end of the century, the study of falcons had expanded beyond falconry and taxonomy. Scholars, like Buffon, began to focus on the morphological and behavioral adaptations of falcons. An intriguing example is M. Huber's text "*Observations sur le vol des oiseaux de proie*" (Huber 1784). This concise monograph includes fascinating illustrations of how birds of prey navigate around fixed points such as cliffs and trees. Additionally, it features comparative drawings of various falcon species, allowing for comparisons of their sizes, plumage patterns, and morphometric traits (Fig. 5.6). Although different from traditional manuals, this text clearly demonstrates new approaches to studying and understanding this group of birds.

Table 5.1 Taxonomy of Western Palearctic falcons before and after Linnaeus '*Systema Naturae*'

Taxon	Gessner 1585	Linnaeus	Location	After Linnaeus	Location
Genus *Falco*		1758			
F. naumanni	–	–	–	Fleischer (1818)	Sicily, Italy
F. tinnunculus	*tinnunculo accipitre*	1758	Sweden	–	–
F. vespertinus	–	1766	St. Petersburg, Russia	–	–
F. eleonorae	–	–	–	Gené (1839)	Sardinia, Italy
F. concolor	–	–	–	Temminck (1825)	Gulf of Aqaba
F. c. columbarius	*aesalone*	1758	South Carolina, USA	–	–
F. c. aesalon	–	–	–	Tunstall (1771)	–
F. c. subaesalon	–	–	–	Brehm (1827)	Iceland
F. subbuteo	*dendrofalco*	1758	Sweden	–	–
F. biarmicus	–	–	–	Temminck (1825)	South Africa
F. b. feldeggii	*falcone cui pedes caerulei*	–		Schlegel (1843)	Dalmatia
F. cherrug	*falcone in genere*	–	–	JE Gray (1834)	India
F. rusticolus	*hierofalcho*	1758	Sweden	–	–
F. peregrinus	*falcone gibboso*	–	–	Tunstall (1771)	Great Britain
F. p. calidus	*falcone peregrinus*	–	–	Latham (1790)	–
F. p. babylonicus?	*falcone nigro*	–	–	Sclater (1861)	–
F. p. pelegrinoides	*falcone rubro*	–	–	Temminck (1829)	–
F. p. brookei	*falcone rubro*	–	–	Sharpe (1873)	–

5.5 Western Palearctic Falcons

5.5.1 *Small Falcons*

During this period, the relationship between humans and falcons underwent a significant transformation. Falcons were no longer valued exclusively for falconry; they began to be perceived as pests. As a result, the Eurasian hobby was classified as a harmful species in Finnish legislation under the Royal Decree on Avian Pest of 1741 (Mykrä et al. 2005). Unfortunately, this approach became common policy in

Fig. 5.6 Comparison table of various perched falcons. The author identified the following: (1) and (2) female gyrfalcons; (3) male gyrfalcon; (5) male peregrine falcon; (7) male common kestrel; (8) Eurasian hobby; and (9) merlin. (Source: ETH-Bibliothek Zürich, permanent link: https://doi.org/10.3931/e-rara-7411)

many European countries during the nineteenth century and persisted well into the latter half of the twentieth century (van Oorscht 1974; Bijleveld 1974). Conversely, the expansion of human activities and the construction of man-made structures created new opportunities for adaptable falcons, providing both prey and suitable nesting sites. For instance, most lesser kestrel colonies in Spain are located in large buildings constructed during the fifteenth to seventeenth centuries, a period when the Spanish Empire was at its peak (Negro et al. 2020).

5.5.2 Large Falcons

The analysis of historical knowledge represents a major achievement in understanding falcons during this period. Efforts were primarily focused on standardizing the nomenclature and classification of falcon species, although progress in understanding their distribution and status across Europe was limited. Nevertheless, records of economic transactions related to falcon trade remain. Notably, in Prussia, continuous falcon trading is documented in administrative records, despite a decrease in

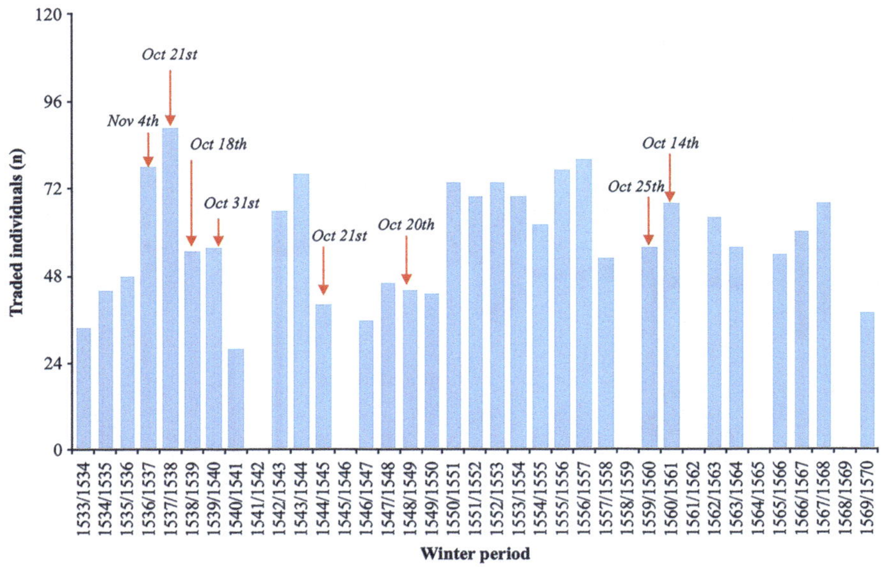

Fig. 5.7 The annual number of falcons (primarily peregrine falcons) traded in Prussia from 1533 to 1569, with red arrows marking the onset of captures. (Data source: Heckmann and Heckmann 2017).

volume from the low Middle Ages (see also Sect. 4.3.2; Heckmann and Heckmann 2017). Since 1558, there has been a noticeable rise in years with low falcon catches, as indicated in "*Wiener Falkenheilkunde*" from the fifteenth century (Fig. 5.7; Heckmann and Heckmann 2017). Significant fluctuations were observed in the years 1535/1536, 1537/1538, 1543/1544, and 1549/1550, while the annual export of falcons between 1557 and 1569 was slightly lower compared to the period from 1536 to 1556 (Fig. 5.7; Heckmann and Heckmann 2017). These fluctuations likely align with the onset of the Little Ice Age in the sixteenth century, marked by global climate changes (Heckmann and Heckmann 2017). The onset of winter around September in the Kola Peninsula and Siberian tundra, combined with windy and cool-wet summer conditions, likely impacted the number of falcon offspring (Anctil et al. 2014; Heckmann and Heckmann 2017).

In Malta, trapped falcons were formally handed over to the Royal Falconer, with official records detailing the species, quantity, and expenses (Fenech 2015). Archival records from Palermo and the National Library of Malta show that between 1431 and 1789, an annual shipment of 4–51 falcons was sent to the Kings of France, Naples, and the King of the Two Sicilies (Fig. 5.8). These falcons were trapped during migration in Malta using vertical nets (Fenech 2010). The identification of lanner falcons (referred to as "*laniers de Sicilie*") was particularly challenging, resulting in low numbers from the late 1600s to the mid-1700s (Fig. 5.8). Although the saker falcon was prevalent before 1600, it experienced a significant decline and disappeared entirely by the early 1700s (Fig. 5.8). Peregrine falcons, categorized by age

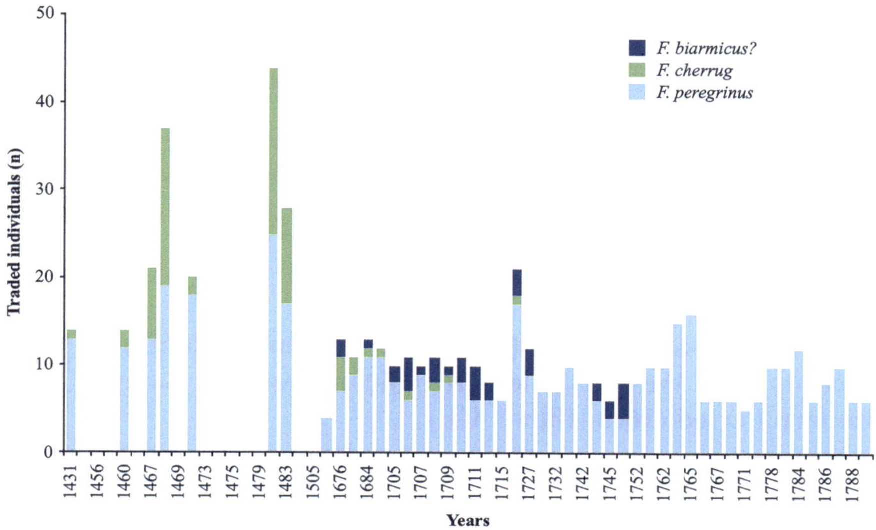

Fig. 5.8 Falcon captures for gifting by Maltese authorities to European rulers. (Fenech 2010, 2015)

and sex as "*haggard*" (adult females), "*gentili*" (*juveniles*), "*peregrini*," and "*terziruoli*" (adult males), continued to be captured after the mid-1700s but also suffered a notable decline, similar to other falcons (Fig. 5.8; Fenech 2010). It is plausible that the onset of the Little Ice Age had a negative impact on falcon species, especially those from Arctic and northern regions. Nonetheless, as late as the mid-eighteenth century, around 50 falcons were still being captured annually on the island of Gozo (Fenech 1992).

5.6 Conclusions

During the Early Modern Period, knowledge about falcons was significantly enhanced by the rediscovery of ancient Greek and Latin texts on natural history. Rather than focusing solely on falconry, a new discipline—ornithology—began to take shape, offering a renewed perspective on the study of birds. Research during this time was primarily divided into two branches: categorizing birds based on their morphological traits and understanding their behaviors in relation to their natural habitats. This era emphasized refining existing knowledge over seeking new evidence, partly due to the decline in the falcon trade.

Several key points can be highlighted:

- The stabilization of falcon taxonomy through the adoption of binomial nomenclature
- Recognition of subspecific variability while the overall number of species remains relatively constant

- Improved accuracy in the depiction of species in illustrations accompanying scientific texts
- Emergence of early studies focusing on the behavior and ecology of species beyond taxonomy and falconry
- A more precise understanding of falcon species distribution across Europe
- Continued maintenance of incorrect knowledge and misidentifications, such as those concerning the lanner falcon
- Inclusion of falcons as harmful species in pest control practices

Appendix: Fossil Remains (*f*), Artifacts (*a*), and Literature Records (*r*) of Western Palearctic Falcons from the Early Modern Period (16th–18th c. CE)

Species	Dates	Location	References
F. naumanni[(a)]	14th–17th c. CE	Andalusia, Spain	Negro et al. (2020)
F. tinnunculus[(f)]	Early 14th–18th c. CE	Koło castle, Poland	Bochenski et al. (2018)
F. tinnunculus[(f)]	13th–18th century	Kraków, Poland	Bochenski et al. (2018)
F. tinnunculus[(f)]	Late 14th–16th century	St Gregory's Priory, Canterbury, UK	Poole (2018)
F. tinnunculus[(f)]	13th–16th century	Felsőnyék-Várhegy, Hungary	Jánossy (1985)
F. tinnunculus[(f)]	15th–16th century	Székelykeresztúr-Udvarház, HU	Gál (2008)
F. tinnunculus[(f)]	15–16th c. CE	Beja, Portugal	Moreno-García and Pimenta (2010)
F. tinnunculus[(r)]	15–17th c. CE	Hague Buitenhof, The Netherlands	Prummel (2018)
F. tinnunculus[(f)]	Mid-17th-early 19th c. CE	Ojców Castle, Poland	Religa-Sobczyk et al. (2023)
F. tinnunculus[(f)]	17–18th c. CE	St Peter's Abbey, Ghent, Belgium	Goddeeris et al. (2002)
F. vespertinus	–	–	
F. eleonorae[(r)]?	16th c. CE (trade)	Kythera, Greece	Arbel (2009)
F. columbarius[(a)]	17th c. CE	Ireland	Grassby (1997)
F. subbuteo[(f)]	Post-medieval	Hague Buitenhof, The Netherlands	Zeiler (2010)
F. subbuteo[(f)]	15th–18th century	Tykocin, Poland	Bochenski et al. (2018)
F. subbuteo[(f)]	16th century	Montsoriu Castle, Spain	Dalmau (2014)
F. biarmicus[(a)]	15th c. CE	Sicily and Apulia, Italy	d'Arcussia (1598)
F. cherrug[(f)]	16–18th c. CE	Sofia, Bulgaria	Boev (2016)

(continued)

Species	Dates	Location	References
F. rusticolus(f)	17th c. CE	Trondheim, Norway	Lie (2018)
F. peregrinus(r)	17th c. CE	Isle of Anglesey, Wales, UK	Ray (1678)
F. peregrinus(r)	16th c. CE	Bologna, Italy	Aldrovandi (1599)
F. peregrinus(f)	Post-medieval	Nonsuch Palace, UK	Ehrlich et al. (2023)
F. peregrinus(r)	15–17th c. CE	Hague Buitenhof, The Netherlands	Holmes (2017)
Falco sp.(f)	14th–16th c. CE	Vilnius Lower Castle, Lithuania	Prummel (2018)

References

Aldrovandi U (1599) Ornithologiae. Franciscum de Franciscis Senensem, Bononiae

Almaça C (1997) Falconers: the first Portuguese naturalists. Arch Nat Hist 24:175–187. https://doi.org/10.3366/anh.1997.24.2.175

Anctil A, Franke A, Bêty J (2014) Heavy rainfall increases nestling mortality of an arctic top predator: experimental evidence and long-term trend in peregrine falcons. Oecologia 174:1033–1043. https://doi.org/10.1007/s00442-013-2800-y

Arbel B (2009) 2009. Venice and Kytherian Falcons: Ecological Aspects of Renaissance Colonialism. In: Diethnes H (ed) Panionio Synedrio, Kythera 21–25 Martiou 2006, 3. Society of Kytherian Studies. Athens. Pp. 38-45

Arcussia (d') C (1598) La fauconnerie de Charles d'Arcussia, seigneur d'Esparron de Paillieres & de Courmes, Gentilhomme Provençal, divisée en trois livres. Jean Tholosan, Aix

Belon P (1555) Histoire de la Nature des Oyseaux. Gillaume Cavellat, Paris

Bijleveld M (1974) Birds of prey in Europe. MacMillan Press, London

Birkhead TR, Charmantier I (2009) History of ornithology. In: Encyclopedia of life sciences (ELS). Wiley, Chichester. https://doi.org/10.1002/9780470015902.a0003096

Bochenski ZM, Tomek T, Wertz K, Wojenka M (2018) Falconry in Poland from a zooarchaeological perspective. In: Gersmann K-H, Grimm O (eds) Raptor and human—falconry and bird symbolism throughout the millennia on a global scale—vol 1/3. Wachholtz, Kiel, pp 1399–1410

Boev Z (2016) Subfossil vertebrate fauna from forum Serdica (Sofia, Bulgaria), 16–18th Century A.D. Acta Zool Bulg 68:415–424

Boyd MJ, Jessop L (1998) A 'truly amiable gentleman': new light on the life and work of Marmaduke Tunstall (1743–1790) of Wycliffe, North Yorkshire. Arch Nat Hist 25:221–235. https://doi.org/10.3366/anh.1998.25.2.221

Buffon (de) Leclerc GL (1785) Histoire naturelle générale et particulière. t. I et II. Imprimerie Royale, Paris

Coureas N (2017) Hunting in Lusignan and Venetian Cyprus (1192–1570). Synthesis 10:103–115. https://doi.org/10.12681/syn.16246

Dalmau VN (2014) Estudi de les pautes d'accés i consum dels recursos animals a partir de l'arqueozoologia. PhD thesis, Universitat Autònoma de Barcelona, Barcelona

De Smet IAR (2018) Princess of the North: perceptions of the gyrfalcon in 16th century western Europe. In: Gersmann K-H, Grimm O (eds) Raptor and human—falconry and bird symbolism throughout the millennia on a global scale—vol 1/4. Wachholtz, Kiel, pp 1543–1569

Egerton FN (2003) A history of the ecological sciences, part 11: emergence of vertebrate zoology during the 1500s. Bull Ecol Soc Am 84:206–212

Ehrlich F, Piličiauskienė G, Blaževičius P (2023) Bird remains from Vilnius Lower Castle, Lithuania (13th–19th centuries) reveal changes in social status and unusual bird pathologies. Int J Oestoarchaeol 33:731–741. https://doi.org/10.1002/oa.3193

Ellenius A (1997) Ulisse Aldrovandi as a source of Olof Rudbeck's Book of Birds. In: Beretta M, Frängsmyr T (eds) Sidereus Nuncius and Stella Polaris: the scientific relations between Italy and Sweden in early modern history. Science History Publications, Canton, pp 47–56

Fenech N (1992) Fatal Flight: the Maltese obsession with killing birds. Quiller Press, London

Fenech N (2010) A complete guide to the birds of Malta. Midsea, Malta

Fenech N (2015) Raptors and falconry in Malta: culture, history and conservation. In: Szymak U, Przemsław S (eds) Falconry—its influence on biodiversity and cultural heritage. Muzeum Podlaskie w Białymstoku, Białymstok, pp 121–168

Gál E (2008) New data to the bird bone artefacts from Hungary and Romania. In: Luik H, Choyke AM, Batey CE, Lõugas L (eds) From hooves to horns, from mollusc to mammoth. Manufacture and use of bone artefacts from prehistoric times to the present. Proceedings of the 4th Meeting of the ICAZ Worked Bone Research Group. Muinasaja teadus 15, pp 325–338

Gassó Miracle ME (2008) The significance of Temminck's work on biogeography: early nineteenth century natural history in Leiden, The Netherlands. J Hist Biol 41:667–716. https://doi.org/10.1007/s10739-008-9158-8

Gessner K (1554) Historiae animalium liber II qui est de avium natura. Froschover, Tiguri

Gessner K (1585) Historia animalium liber III, qui est de avium natura. Roberti Cambieri, Frankfurt

Goddeeris B, De Smet G, Roggeman W (2002) Some gastronomic aspects of bird species in still life paintings of Frans Snyders (Antwerp, 1579–1657). Revue belge de philologie et d'histoire 80:1431–1448

Grassby R (1997) The decline of falconry in Early Modern England. Past Present 157:37–62

Hatch RA (2000) Correspondence networks. In: Applebaum W (ed) Encyclopedia of the scientific revolution from Copernicus to Newton. Garland, New York, pp 168–170

Heckmann M-L, Heckmann D (2017) Falken im Ordensland und Herzogtum Preußen—ein ökologisch und ökonomisch knappes Gut. Preußenland 8:7–40

Holmes M (2017) Southern England: a review of animal remains from Saxon, Medieval and Post-Medieval archaeological sites. Historic England, Research Report Series 8

Holterman B (2020) The Fish Land. De Gruyter, Oldenbourg

Jánossy D (1985) Wildvogelreste aus archäologischen Grabungen in Ungarn (Neolithicum bis Mittelalter). Fragm Mineralogica et Palaeontol 12:67–103

Latham J (1790) Index Ornithologicus, sive systema ornithologiæ: complectens Avium divisionem in classes, ordines, genera, species, ipsarumque varietates, vol 1. Leigh & Sotherby, London

Lie RO (2018) Falconry, falcon-catching and the role of birds of prey in trade and alliance gifts in Norway (800–1800 AD) with an emphasis on Norwegian and later foreign participants in falcon catching. In: Gersmann K-H, Grimm O (eds) Raptor and human—falconry and bird symbolism throughout the millennia on a global scale—vol 1/2. Wachholtz, Kiel, pp 727–786

Linneaus C (1758) Systema naturae per regna tria naturae: secundum classes, ordines, genera, species, cum characteribus, differentiis, synonymis, locis, 10th edn. Laurentius Salvius, Stockholm

Lorenzi L, Manetti S, Vanni V (1768) Storia naturale degli uccelli, vol 1. Stamperia Mouckiana, Florence

Moreno-García M, Pimenta C (2010) Beyond chicken: avian biodiversity in a Portuguese late medieval urban site. Groningen Archaeol Stud 12:261–275

Mykrä S, Vuorisalo T, Pohja-Mykrä M (2005) A history of organised persecution and conservation of wildlife: species categorizations in Finnish legislation from medieval times to 1923. Oryx 39:275–283. https://doi.org/10.1017/S0030605305000797

Naumann JA, Naumann JF, Hennicke CR (1899) Naturgeschichte der Vögel Mitteleuropas, vol 5. Köhler, Gera-Untermhaus

Negro JJ, Prenda J, Ferrero JJ, Rodríguez A, Reig-Ferrer A (2020) A timeline for the urbanization of wild birds: the case of the lesser kestrel. Quat Sci Rev 249:106638. https://doi.org/10.1016/j.quascirev.2020.106638

Olson RJM, Mazzitelli A (2007) The discovery of a cache of over 200 sixteenth-century avian watercolours: a missing chapter in the history of ornithological illustration. Master Drawings 45:435–521

Perfetti S (2000) Aristotle's zoology and its Renaissance commentators (1521–1601). Leuven University Press, Leuven

Petersen A (1998) Icelandic birds. Vaka-Helgafell, Reykjavík

Poole K (2018) Zooarchaeological evidence for falconry in England, up to AD 1500. In: Gersmann K-H, Grimm O (eds) Raptor and human—falconry and bird symbolism throughout the millennia on a global scale—vol 1/3. Wachholtz, Kiel, pp 1027–1053

Prommer M, Bagyura J, Chavko J, Uhrin M (2012) Migratory movements of Central and Eastern European Saker falcons (*Falco cherrug*) from juvenile dispersal to adulthood. Aquila 119:111–134

Prummel W (2018) The archaeological-archaeozoological identification of falconry—methodological remarks and some Dutch examples. In: Gersmann K-H, Grimm O (eds) Raptor and human—falconry and bird symbolism throughout the millennia on a global scale—vol 1/2. Wachholtz, Kiel, pp 467–478

Ray J (1678) The ornithology of Francis Willoughby. John Martyn, London

Raye L (2018) Robert Sibbald's Scotia Illustrata (1684): a faunal baseline for Britain. Notes Rec 72:383–405. https://doi.org/10.1098/rsnr.2017.0042

Religa-Sobczyk J, Wertz K, Lõugas L, Wojenka M, Lemanik A, Wojtal P (2023) Zooarchaeological evidence from Medieval Ojców Castle, Lesser Poland. Heritage 6:258–283. https://doi.org/10.3390/heritage6010013

Tunstall M (1771) Ornithologia Britannica: seu avium omnium britannicarum tam terrestrium quam aquaticarum catalogus. Published Privately, London

van de Wall JWM (2004) The Loo Falconry: The Royal Loo Hawking Club 1839–1855. Hancock House Publishers, British Columbia/Washington

van Oorscht JMP (1974) Vorstelijke vliegers en valkenswaardse valkeniers sedert de seventiende eeuw. Stichting Zuidelijk Historisch Contact, Tilburg

Zeiler JT (2010) Hunting the hunters: owls and birds of prey as part of the falconers' game bag. In: Prummel W, Zeiler JC, Brinkhuizen BC (eds) Birds in archaeology Groningen Archaeological Studies 12. Barkhuis, Groningen, pp 163–168

Chapter 6
Late Modern Period

6.1 Introduction

During the Early Modern Period, naturalists laid the foundation for the modern taxonomy and classification of falcon species (see Table 5.1). New exploration voyages and trade with distant lands provided opportunities for museums and collectors to acquire a variety of natural curiosities (Farber 1977; Morris 1993). However, birds were often inadequately represented due to the challenges of preserving them in recognizable conditions (Farber 1977: Morris 1993). Preserving birds was necessary not only to prevent decay but also to facilitate taxidermy for accurate comparison and classification (Morris 1993). Additionally, with the decline of the flourishing falcon trade, it became nearly impossible for ornithologists to acquire live specimens for examination (Naumann et al. 1899). Starting from the mid-1800s, overcoming technical challenges allowed for the collection of a vast number of samples from each species (Morris 1993). Ultimately, the influx of new specimens significantly influenced the development of falcon taxonomy and systematics in Europe (Gassó Miracle 2008). At the Great Exhibition of 1851, the taxidermist John Hancock presented a naturalized gyrfalcon capturing a gray heron in a simulated environment. During the subsequent commercial boom, many specialized businesses emerged to meet the growing demands of collectors. These were not just museums but also wealthy clients competing for extensive collections of both local and exotic specimens. Exoticism was not limited to species from distant continents but also included those from within Europe. For example, the lanner falcon, often considered rare, was highly prized (Orlando 1957). Over 23 years, about 400 lanner falcons were captured in southern Italy, peaking at 43 between 1927 and 1928 (Arrigoni degli Oddi 1929).

This market continued to grow steadily, especially as falcons were officially recognized as harmful species. This notion, already present by the end of the previous century, was reinforced by local laws and implemented by private citizens and

G. Leonardi, *The History of Western Palearctic Falcons*,
https://doi.org/10.1007/978-3-031-86528-2_6

gamekeepers (Morris 1993). With the rise of small-game management in the nineteenth century, persecution of raptors peaked and extended to species of various sizes (Newton 1979). For instance, a single estate in Scotland documented the destruction of 1372 birds of prey in just 3 years (1837–1840), while a total of 39,233 raptors were recorded as having been slaughtered throughout the Netherlands between 1852 and 1857 (Pearsall 1950; Braaksma et al. 1959). However, there is limited scientific evidence available on the effects of this persecution, as most declines in numbers occurred between 1850 and 1900, before biologists began systematically recording them (Newton 1979).

At the end of the nineteenth century, a significant shift in the approach to studying falcons was imminent (see Box 6.1). While taxonomy and systematics remained essential, manuals increasingly featured field observations on the reproductive biology of each species, along with detailed descriptions of various behavioral patterns, both within and between species. However, for scholars of the time, the primary sources of information were the locally captured individuals preserved in museums and private collections. Endless debates and a multitude of taxa emerged from meticulous analyses, often based on just a handful of specimens (see Sect. 6.2.1). In discussing the supposed morphometric differences between gyrfalcons from Norway and Iceland, J. Gould points out the methodological challenges: "*The question of whether there are two species has been raised by systematic ornithologists. However, we are uncertain whether the specimens in our various collections provide enough data to support a clear separation*" (Gould 1837). On the other hand, these captures provided a definitive reference point for the previously uncertain native locations of individual species (see Sect. 6.4). The act of capturing and subsequent taxidermy served as tangible, verifiable evidence within the scientific community to support their claims. Following this straightforward logic, ornithologists assessed species abundance in the field by the number of captures. German naturalists who travelled primarily through Tunisia, Morocco, and Algeria systematically collected parents, eggs, and *juveniles* from all lanner falcon nests they encountered during their expeditions (Koenig 1892; Erlanger 1898; Kleinschmidt 1901). Consequently, they concluded that the lanner falcon was the most prevalent raptor in these regions. While this assumption carries certain biases, the comprehensive analysis of museum specimens in Sect. 6.4 focuses exclusively on those collected during the breeding season, aiming to capture perspectives from the Late Modern Period.

The Late Modern Period, spanning from 1800 to 1945, marked the beginning of contemporary history. Despite lasting just a century and a half, human activities significantly impacted the distribution and persistence of falcon populations. Since the time of the Neanderthals (see Sect. 1.5), the exploitation of falcons has notably increased, particularly during Ancient Times for ceremonial purposes (see Box 3.3) and within the organized market for falconry (Fig. 6.1; see Sect. 4.2.3).

RICERCA DI FALCHI VIVI

Con lo scopo di tentarne l'allevamento ad uso di Sport (Falconeria) si desidera avere vivi falchi delle seguenti specie:

Falcone (*Falco peregrinus*) di tutte le età.

Astore (*Astur palumbarius*) giovane prima della muta

Sparviere (*Accipiter nisus*) femmina e giovane.

Smeriglio (*Aesalon regulus*) di tutte le età.

Chi ne avesse o potesse procurarne è pregato darne avviso alla amministrazione di questo giornale in **Siena.**

Fig. 6.1 An announcement in a 1900 issue of the Italian journal "*Avicula*" seeks live falcons and hawks to initiate captive breeding for falconry purposes. (Source: BHL https://www.biodiversitylibrary.org)

6.2 The Nineteenth Century

6.2.1 Taxonomy

Given the fluidity of zoological nomenclature in the early nineteenth century, the number of taxa grew exponentially as genera and species were split based on meristic morphological traits (Gassó Miracle 2008). Among the numerous binomial names, the final species of falcons not classified by Linnaeus in 1758 were identified (see also Table 5.1). The species yet to be defined included rare ones (lanner and sooty falcons), easily misidentified species (saker falcon and lesser kestrel), those exhibiting various morphs (Eleonora's and gyrfalcons), and those encompassing different yet closely related subspecies (peregrine falcons).

The lesser kestrel was recognized as a distinct species from the common kestrel by Johann G. Fleischer in 1818 naming it *F. naumanni* in honor of his friend Johann F. Naumann (Laurop and Fischer 1818). However, in his work "*Naturgeschichte der Vögel Deutschlands*," Naumann identified the lesser kestrel as *F. cenchris*, a name that already coexisted with *F. tinnunculoides*, given by C. J. Temminck.

C. Giuseppe René first introduced the Eleonora's falcon species during a meeting on March 3, 1829, at the Royal Academy of Turin, presenting a specimen captured in Sardinia (Anonymous 1893). The full description, with the detailed observations, was published the following year (Gené 1840). The existence of a dark morph in this species led to confusion with the closely related sooty falcon (*F. concolor*), which inhabits the Red Sea and the East African coast (Warthausen 1860; Verreaux and des Murs 1862). This confusion was clarified by T. Krueper in 1862, who, rather than relying on museum specimens, spent considerable time observing nesting colonies on the island of Naxos (Krueper 1862, 1864). His detailed account of the species' breeding season remained a key contribution to the field until at least the 60s of the twentieth century (Vaurie 1961).

The sooty falcon syntypes were collected by E. Rüppel on the island of Barqan (Gulf of Aqaba) and are currently preserved at the National Museum of Natural History of Leiden (Rüppel 1845). Since C. J. Temminck's initial description of the species in 1825, confusion has arisen with two other falcon species sharing similar physical traits: Eleonora's Falcon (mentioned earlier) and the gray kestrel (*F. ardosiaceus*), which were mistakenly illustrated alongside Temminck's description (Temminck 1825).

In 1827, C.L. Brehm added the subspecies *F. c. subaesalon* to the merlin group, identifying it as a resident in Iceland and a migrant in northern Europe (Brehm 1827). Defining the lanner falcon as a species (*F. biarmicus*), along with its presumed subspecies in the Western Palearctic, has proven to be considerably more complex (Leonardi and Sutton 2020). In many ways, the definition of this species exemplifies the methodological approach of nineteenth-century ornithologists. Beginning with Linnaeus, who described the species based on a young gyrfalcon, the lanner falcon (referred to as "*blue legs*") appeared to be widespread throughout Europe (see Sect. 4.4.2 for further details). With the rapid growth of collections of naturalized individuals, it soon became paradoxical that, despite the species' presumed wide distribution, museums and private collections lacked any specimens (Gould 1837). Accordingly, the nonexistence of the lanner falcon as a species separate from the peregrine falcon was proposed (Savi 1827; Benoit 1840; Malherbe 1843). However, modern ornithologists remained intrigued by the influence of past falconry knowledge. Initially thought to be a barbary falcon by Johann Natter (1787–1843), a specimen collected from Dalmatia was later classified as a new species by Schlegel (1844). He named it *Falco feldeggii*, later renaming it *Falco lanarius*. This change was believed to represent the "*true*" lanner falcon described by Belon du Mans (1555) and recognized by old falconers (Schlegel 1844; Schlegel and Verster van Wulverhorst 1845–1853). Similar to other monomorphic but clearly polytypic species, the approach in the early 1900s was to consider all variants as subspecies (Hartert 1912; Leonardi 2001). R. Bowdler Sharpe (1873) described a new subspecies of peregrine falcon based on two specimens collected in Sardinia, Italy. Interestingly, besides the usual descriptions of plumage patterns to highlight differences, he also added key measurements in a comparative table (Sharpe 1873). Although a modern approach, it is still insufficient due to the small sample size and lack of significant statistical differences (Fig. 6.2).

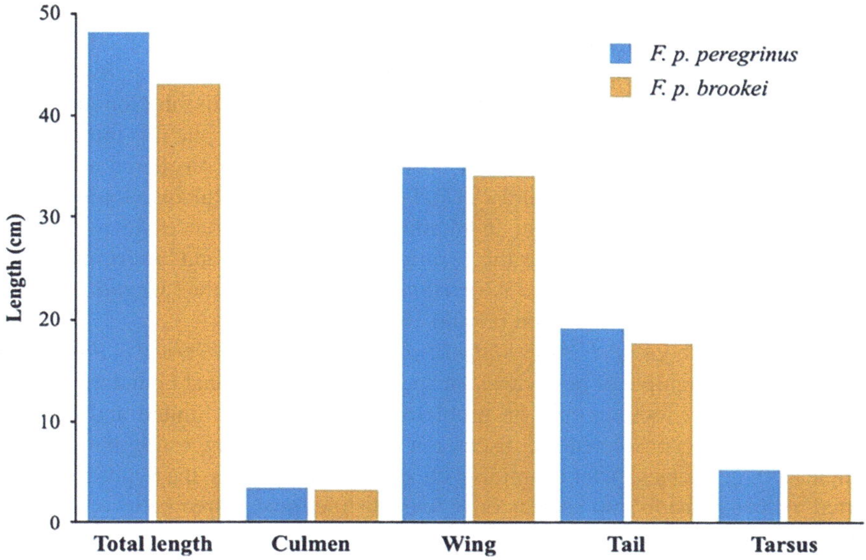

Fig. 6.2 Differences in key morphometric measurements between *F. p. peregrinus* and *F. b. brookei* (then considered a separate species). Measurements were taken from skins, not mounted specimens. (Data source: Sharpe 1873)

6.2.2 Ornithological Literature

In the early 1800s, ornithologists had mixed knowledge about falcons, inherited from earlier times. John Hunt, an ornithologist active from the late 1700s to the mid-1800s, used different names for the peregrine falcon (common falcon, blue-backed falcon, gentil falcon, haggard falcon, slight falcon) and the merlin (stone falcon) in his book "*British Ornithology*" (Hunt 1818). He argued that all these names for the peregrine falcon actually refer to the same species, *Falco gentilis* by Linnaeus, with differences only in age or capture circumstances (Hunt 1818). Additionally, the differences in appearance between males and females in species like the merlin and the common kestrel were well understood (Hunt 1818). The arrival of many birds from English, French, and Dutch colonies helped ornithologists better understand the *Falco* genus by expanding the number of species within it. In his influential work "*The Animal Kingdom*," Georges Cuvier includes numerous African, Asian, and American species (Cuvier 1829). This initiated a comparative process that later helped organize the main species and their subspecies. For instance, C. J. Temminck's classification of southern African populations of the lanner falcon forms the basis for classifying the Western Palearctic subspecies today (Hartert 1912; Leonardi 2001).

John Gould's "*The Birds of Europe*" is a significant work from the mid-1800s by the renowned Victorian era English ornithologist (Gould 1837). Gould regarded the gyrfalcon as native to most of Northern Europe, with occasional visits to the Orkney and Shetland Isles (Gould 1837). He questioned whether it should be divided into multiple species across different territories but ultimately viewed them as one (Gould 1837). Concerning the lanner falcon, Gould emphasizes the scarcity of specimens and primarily describes the saker falcon, detailing its distribution and morphology, with an accompanying illustration (Gould 1837). Despite this initial error, Gould correctly deduces that the lanner falcon should not be considered a *juvenile* form of the peregrine falcon (Gould 1837). For the red-footed falcon, Gould clearly described the sexual dimorphism in plumage and the differences in immatures (Gould 1837). He recognized it as the "*orange-legged hobby*" and "*Ingrian falcon*" from Russia and also as the *Falco vespertinus* of Gmelin and the "*singular variety of the hobby*" described by Buffon (Gould 1837).

Charles Bree's book "*A History of the Birds of Europe Not Observed in the British Isles*" is a good example of how views on species crossed national boundaries (Bree 1859). Actually, it focuses only on problematic species with limited and debated information. He reconsidered the discussion on gyrfalcons, suggesting there might be at least three species. After examining hundreds of specimens, the hypothesis supported a clear division into species or at least distinct races. Bree believed that the true "*ger-falcon*" is the one found in Norway, where no white or whitish birds have ever been seen (Bree 1859). For the saker falcon, he agrees with H. Schlegel's view that it is a distinct species from the lanner falcon (Bree 1859). He also cited Schlegel regarding the lanner falcon, crediting him for "*applying the right name to the right bird*" (Bree 1859). Although there was still confusion about this species, a critical analysis was beginning to clarify its distribution in Europe (see Sect. 6.4.8; Bree 1859).

By the end of the century, H. E. Dresser, in his book *The Birds of Europe*, broadened the geographical range to include the entire Western Palearctic (Dresser 1881). Like earlier manuals, significant attention is given to the presumed gyrfalcon species in Scandinavia, Iceland, and Greenland. Based on data from non-European colonies, he describes the peregrine falcon as the most widespread species of the genus *Falco*, though it appears to be declining in Great Britain. He also highlights the species' long-term use of the same sites, citing nests in northern Finland, and reports a case of recruitment in a pair in England within just two-and-a-half days. Regarding small falcons, he reports that gamekeepers prevent the Eurasian hobby from becoming a settled resident by shooting them. Interestingly, he stated: "*the present generation will, we fear, live to find the hobby classed as one of the rarer British birds.*" The *modus operandi* of the gamekeepers is further illustrated by the systematic destruction of two merlin sites in Great Britain (Table 6.1).

Table 6.1 Persecution of the merlin at two sites in Great Britain, based on reports by Dresser (1881)

Breeding site	1869	1870	1871	1872	1873
Stimes	Parents and brood destroyed	Male shot, female trapped, eggs removed	Eggs removed	No nest	Parents destroyed; eggs removed
Ashopton	Parents and fledglings destroyed	Female, male, and a new recruited female destroyed	Parents and brood destroyed	No nest	–

Fig. 6.3 Lanner falcon eggs collected in 1863 in Egypt: (1) from the Pyramid of Dashoor and (2) from the Pyramid of Giza. (Cochrane 1864; Source: BHL https://www.biodiversitylibrary.org)

An intense trade for eggs (primarily whole clutches) emerged among a large community of collectors, similar to the trade for naturalized individuals (Malan 1848). Hartert (1912), in his description of the color patterns of peregrine falcon eggs, emphasizes that "(…) *they are extremely variable and, unfortunately, a highly popular object for many collectors*." Of course, exotic eggs were the most highly sought after. Accordingly, the first comprehensive paper devoted entirely to lanner falcon eggs was published in one of the early issues of the journal *Ibis* (Cochrane 1864; Fig. 6.3). Unfortunately, the tragic consequence is that a huge number of eggs and clutches were taken worldwide between 1890 and 1930 (over 40,000 per decade; Marini et al. 2020).

Interestingly, the text provides information on prey composition unrelated to falconry practice. These new data come from stomach content examinations of shot individuals. For instance, the study of two female saker falcons revealed remains consisting solely of small mammals.

Produced at the end of the nineteenth century, this text reflects the evolving field of ornithology. In addition to citations from manuals (mainly through the authors), it includes numerous observations from specialized scientific journals. This shift, which began in the latter half of the century, was driven by publications such as *Journal für Ornithologie* (started in 1853) and *Ibis* (1859).

6.3 The Twentieth Century (Up to 1945)

Ernst Hartert (1859–1933) made a significant impact in the early twentieth century by rationalizing and systematizing existing ornithological knowledge, particularly in taxonomy. In 1912 (published in 1922), he authored his comprehensive work "*Die Vögel der paläarktischen Fauna: systematische Übersicht der in Europa, Nord-Asien und der Mittelmeerregion vorkommenden Vögel*" (*The Birds of the Palearctic Fauna: A Systematic Overview of the Birds Found in Europe, Northern Asia, and the Mediterranean Region*). Later, in 1915, he published a paper titled

Box 6.1: ***Naturgeschichte der Vögel Mitteleuropas***

This extensive 12-volume work is a new edition of "*Naturgeschichte der Vögel Deutschlands*" by J. A. Naumann, updated and expanded by his son Johann Friedrich Naumann between 1820 and 1860. The third edition, completed between 1897 and 1905, includes Carl R. Hennicke as one of the authors (Naumann et al. 1899). Additionally, Otto Kleinschmidt, a foremost expert on large falcons (especially hierofalcons), played a significant role in supervising the falcon section (Fig. 6.4; Naumann et al. 1899). The layout and organization of this book served as a model for other contemporary handbooks, such as those by Glutz et al. (1971) and Cramp and Simmons (1980). Accordingly, the paragraphs are organized scientifically for the first time, reflecting the biology of the species. They cover morphology, habitat, reproduction, hunting behavior, and its harmful impact. Despite the updates, numerous notes, insights, and assessments derived directly from falconry literature are still included. Each species account focuses primarily on morphology and taxonomy, with only brief mentions of ecological and behavioral aspects, suggesting a lack of field observation. Conversely, the role of falcons as pests is emphasized in the aforementioned dedicated paragraphs. Although it was generally assumed that these species would be considered pests, the damage they caused was mitigated by their abundance in nature. As a result, a rare species like the gyrfalcon "*can naturally not be considered harmful.*"

In the most recent update, the authors note significant changes since Naumann's initial version, particularly regarding the gyrfalcon. The primary change, aligning with the prevailing research trend, is the subdivision of the species into multiple subspecies. Each species account aimed to define a specific range for its subspecies, with the "*habitat*" section describing both the subspecies and their locations. The large number of subspecies and varying naming conventions resulted in many overlapping territorial areas. At the same time, some subspecies were merged when specimens from different regions showed clear similarities.

Fig. 6.4 Illustrations of adult and *juvenile* gyrfalcons and sakers, the latter called *Falco lanarius*. (Naumann 1820; Source: BHL https://www.biodiversitylibrary.org)

Notes on Falcons in the journal *Novitates Zoologicae*, which he co-founded. Unlike previous manuals, his writings are notably more concise and focused, yet they incorporate extensive morphometric measurements (Hartert 1912, 1915). This was made possible by his role as the ornithological curator of Rothschild's private collection, now held at the Natural History Museum in Tring.

Following Hartert's path by the early 1930s, the systematics and taxonomy of falcons were still shaped by the extensive proliferation of subspecies that had predominantly occurred during the nineteenth century (Table 6.2; Peters 1931). In the contemporary period, a significant number of these taxa have been dismissed as part of an ongoing revision process that has yet to be completed (see Chap. 7 and Table 6.2). Eleonora's falcon and the saker falcon, identified in the first half of the nineteenth century, are accepted along with fewer subspecies than originally listed. Excluded subspecies were regarded as synonyms, either because they were confined to small areas not separate from the main population or because their distribution was unclear (Peters 1931).

In the early twentieth century, papers entirely devoted to falcons in Europe were far fewer than those focused on North American populations. Despite this limited output, notable developments emerged, such as a greater focus on behavioral aspects and a more integrative approach. These studies also began to include photographs taken in natural settings, depicting scenes like falcon pairs caring for their chicks (Gilbert 1927; Ingram and Salmon 1929; Scholze 1933). The rise of ethology inspired a new way of studying falcons, focusing on their behavioral patterns and their ontological and biological significance (Lorenz 1935). Niko Tinbergen (1907–1988), a pioneer in this field, noted a lack of information about the reproductive biology and diet of species such as the Eurasian hobby, relying mostly on occasional observations (Tinbergen 1932). Through careful observation, Tinbergen found that "*the Eurasian hobby male provides the female with food long before egg-laying begins*" (Tinbergen 1932). Indeed, the key contribution to this new approach was the collection of measurable, comparable data that could be analyzed using mathematical tools, such as graphs (Tinbergen 1940).

For instance, Luuk Tinbergen's research on parental care in the common kestrel laid the groundwork for modern behavioral ecology studies focused on falcons (Tinbergen 1940).

Even in this century, the trade in falcons has persisted, mainly as a status symbol and a reflection of a glorious past. One example is Hermann Göring, Adolf Hitler's chief lieutenant, who imported gyrfalcons from West Greenland in 1938 and Iceland in 1940 (Allsen 2006; Barr 2012). However, falcons are still viewed as harmful species while also being exploited for various other purposes. Walter (1979) noted that before the 1940s, Aegean fishermen at the offshore islets of Crete harvested large numbers of Eleonora's falcons, primarily for food. On the other hand, early conservation efforts began, like the Finnish government's Nature Conservation Act of 1923, which fully protected the common kestrel, Eurasian hobby, and red-footed falcon (Mykrä et al. 2005). In Switzerland, the sparrowhawk, goshawk, and

Table 6.2 Systematic arrangements of species and subspecies within the genus *Falco* according to Peters (1931)

Genus	Species	Western Palearctic subspecies
Falco (Linneaus 1758)	*Falco naumanni* (Fleischer 1818) *	
	Falco tinnunculus (Linneaus 1758)	*F. t. tinnunculus* (Linneaus 1758)
		F. t. canariensis (Koenig 1890) *
		F. t. dacotiae (Hartert 1913) *
		F. t. neglectus (Schlegel 1873) *
		F. t. rupicolaeformis (Brehm 1855) *
	Falco vespertinus (Linneaus 1766)	–
	Falco eleonorae (Géné 1839) *	–
	Falco concolor (Temminck 1825) *	–
	Falco columbarius (Linneaus 1758)	*F. c. columbarius* (Linneaus 1758)
		F. c. subaesalon (Brehm 1827) *
		F. c. aesalon (Tunstall 1771)
		F. c. regulus (Pallas 1773) †
	Falco subbuteo (Linnaeus 1758)	*F. s. subbuteo* (Linnaeus 1758)
		F. s. distinguendus (Portenko 1930) *†
		F. s. jugurtha (Hartert and Neumann 1907) *†
	Falco biarmicus (Temminck 1825) *	*F. b. biarmicus* (Temminck 1825) *
		F. b. erlangeri (Kleinschmidt 1901) *
		F. b. feldeggii (Schlegel 1843) *
		F. b. tanypterus (Schlegel 1843) *
		F. b. abyssinicus (Naumann 1904) *
	Falco cherrug (Gray 1834) *	*F. c. cherrug* (Gray 1834) *
		F. c. progressus (Stegmann 1925) *†
		F. c. milvipes (Jerdon 1871) *
	Falco rusticolus (Linneaus 1758)	*F. r. rusticolus* (Linneaus 1758)

		F. r. candicans (Gmelin 1788) †
		F. r. islandus (Brünnich 1764) †
		F. r. uralensis (Severzov and Menzbier 1922) *†
		F. r. obsoletus (Gmelin 1788) †
	F. peregrinus (Tunstall 1771)	*F. p. peregrinus* (Tunstall 1771)
		F. p. calidus (Latham 1790)
		F. p. brookei (Sharpe 1873) *
		F. p. pelegrinoides (Temminck 1829) *

† = dismissed subspecies; * = subspecies described during the nineteenth and twentieth centuries

peregrine falcon were the birds of prey that could be killed (Nice 1933). Bounties were once offered for these species, but this was stopped when it was found that beneficial species were also being killed (Nice 1933).

6.4 Western Palearctic Falcons

As previously mentioned, specimen collection in the field was conducted year-round across all European countries and their colonies. Traditional methods predominantly involved capturing falcons (mainly *juveniles* and sub-adults, but also adults) during their seasonal movements to maximize yield (see Sect. 5.5.2). Conversely, capturing falcons during the breeding season, especially large species, proved significantly more challenging due to the variability of nest sites, leading to a noticeable bias in capture numbers (Fig. 6.5). Additionally, species from remote or difficult-to-access regions, such as the sooty falcon, represent less than 1% of specimens. Consequently, knowledge of breeding ranges was often inadequate and frequently inaccurate. Predictably, species with broad distributions, such as the common kestrel, Eurasian hobby, and peregrine falcon, show a much smaller bias (Fig. 6.5). Furthermore, capture locations have a significant impact on these percentages (see individual species accounts; Fig. 6.6). For example, established trade practices have resulted in the majority of gyrfalcons in collections coming almost exclusively from Iceland (see Sect. 6.4.10). German travellers such as C. von Erlanger and Alexander Koenig collected several adults, eggs, and chicks of the

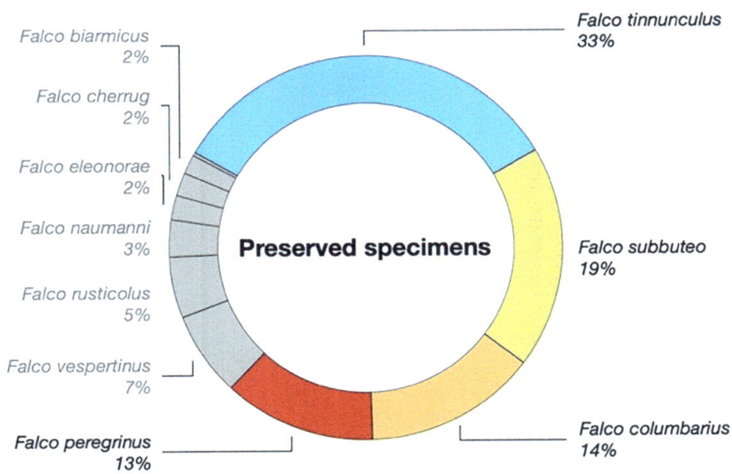

Fig. 6.5 Percentage of specimens from various falcon species in leading Natural History Museums during the Late Modern Period. The dataset includes preserved specimens from 1800 to 1946, specifically from the breeding season months for each species. (n = 4216; Source: Global Biodiversity Information Facility—GBIF, http://www.gbif.org/)

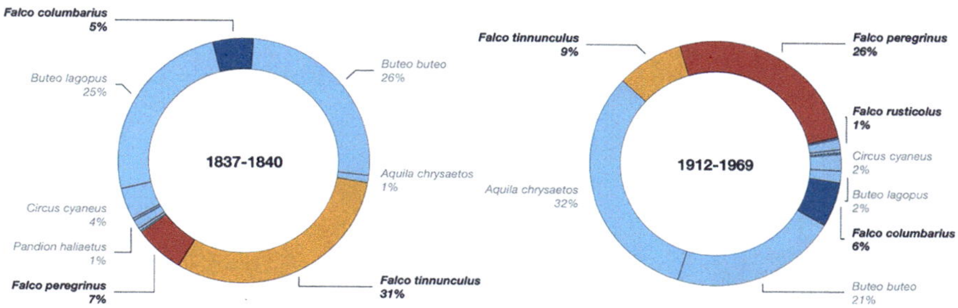

Fig. 6.6 Killing rates of raptors in Scotland recorded as percentages during two different time windows of the Late Modern Period. (Data from Yalden and Albarella 2009)

lanner falcon during their expeditions in Tunisia and Algeria (Elzen 2010). At the beginning of the twentieth century, museums and private collections held an impressive number of specimens from the *erlangeri* race, including numerous adult and immature birds and at least 40 eggs—far surpassing the representation of other subspecies. Systematically collecting broods, shooting both parents, and, where possible, attempting to rear chicks found at a later stage were common aspects of the field approach. Therefore, it is not surprising that R. Gray, in his late nineteenth-century book on the birds of Western Scotland, highlighted the peregrine falcon's resilience despite enduring "*an extraordinary amount of persecution*" (Gray 1871). Despite these numerous sources of bias, the data still offer a valuable overview of the status of falcons in the Western Palearctic.

6.4.1 Falco naumanni

This species, clearly distinct from the common kestrel, was regarded as common in the southern parts of continental Europe, particularly in Spain, Italy, and the south of France (Fig. 6.7; Gould 1837). Gould, while not entirely dismissing the hunting of rodents, notes that it mainly feeds on larger beetles and ants, which it catches in flight (Gould 1837). Bree (1859) reported the lesser kestrel nests in old ruins or the crevices of mountain rocks, particularly in Sicily and the vicinity of Gibraltar, as noted by C. J. Temminck. Considered abundant in Morocco, Algeria, Tunisia, Spain, Sardinia, Sicily, southern and central Italy, and south-eastern Europe, it is also described as a migratory bird that overwinters in tropical and southern Africa, including central Africa (Hartert 1912). It was also noted that, in contrast to the common kestrel, which is usually seen in small, loose groups and whose nests, although often close together, are never colonial, the lesser kestrel is very social, breeding in colonies and migrating in often large flocks (Hartert 1912). It was finally clarified that its diet consists of insects, mainly grasshoppers and beetles, as well as reptiles and small mammals (Hartert 1912).

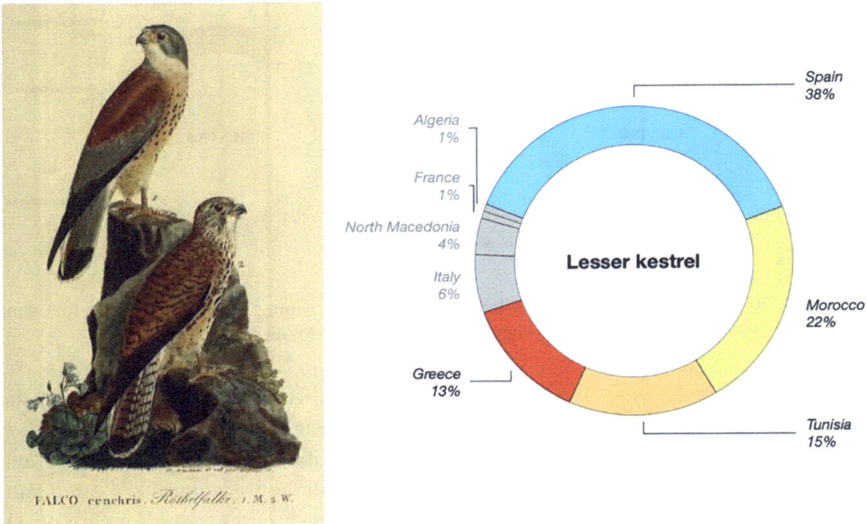

Fig. 6.7 Male and female lesser kestrel *F. cenchris* (= *naumanni*). (Naumann 1820; Source: BHL https://www.biodiversitylibrary.org)—Percentage of lesser kestrel specimens from the Late Modern Period conserved in major Natural History Museums. (Source: GBIF, http://www.gbif.org/)

6.4.2 Falco tinnunculus

This falcon was widely regarded as the most common species among those inhabiting Europe (Fig. 6.8; Fleming 1828; Gould 1837). Fleming (1828) describes it as nesting in tree cavities, rock crevices, and ruins, hunting small mammals and insects through hovering flights. Hartert (1912) offers a highly detailed account of the species' distribution and migratory patterns. It was considered a breeder in Europe from 68° in Scandinavia and 61° in Russia to the Mediterranean, its islands, and North Africa up to the northern Sahara, as well as in North and West Asia. In Northern Europe, Northern Asia, and also in Northern Scotland, the Shetland Islands, and Germany, it migrates. In England, Southern Europe, as well as in the milder regions of Germany and North Africa, it remains all year round. In cities, it sometimes even overwinters in the coldest parts of Germany. Hartert (1912) also notes that its habitat is diverse, preferring open landscapes over forests, though it does not completely avoid them. It nests in many places, including cliff walls, riverbanks, rocks, old buildings, towers, gunports, and nests of other birds, as well as under bridges, in tree cavities, and sometimes on the ground.

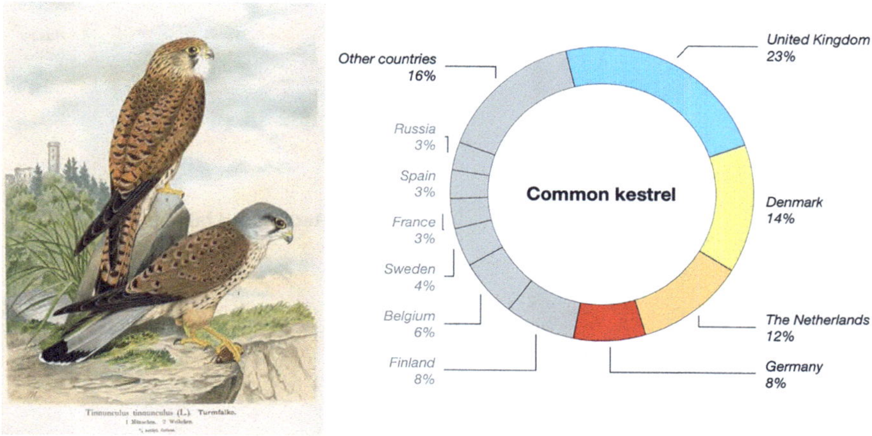

Fig. 6.8 Male and female common kestrel. (Naumann et al. 1899; Source: BHL https://www.biodiversitylibrary.org). Percentage of common kestrel specimens from the Late Modern Period conserved in major Natural History Museums. (Source: GBIF, http://www.gbif.org/)

6.4.3 Falco vespertinus

This species, despite being known by different names, has been identified (accounting for sexual dimorphism and the juvenile stage) through both museum and private collections (Fig. 6.9; Gould 1837). Hartert (1912) considered it a breeding bird in Eastern Europe and West Siberia, extending in Russia from the south to the north up to Arkhangelsk, and in the west as far as the Baltic provinces and Poland. It was commonly found in south-eastern Romania and the plains of Hungary. During migration, it appeared regularly, though irregularly distributed, in the historical regions of East Prussia and Silesia (then part of Russia, Poland, Lithuania, the Czech Republic, and Germany). Occasional sightings were recorded further west, including approximately 40 in the British Isles, as well as in France, Spain, and Northwest Africa. In Tunisia, it was almost a regular spring migrant, while in Italy, it was not uncommon during migration. Hartert (1912) described it as a resident of the plains, nesting in field woodlands, at forest edges, and in open woods. Its diet mainly consisted of insects, especially orthopterans and coleopterans, but also included reptiles and amphibians. For breeding, it often used crow and magpie nests, especially those of the rook, with several nests sometimes in one tree. Nests were sometimes found in groups, and social behavior was also seen during migration.

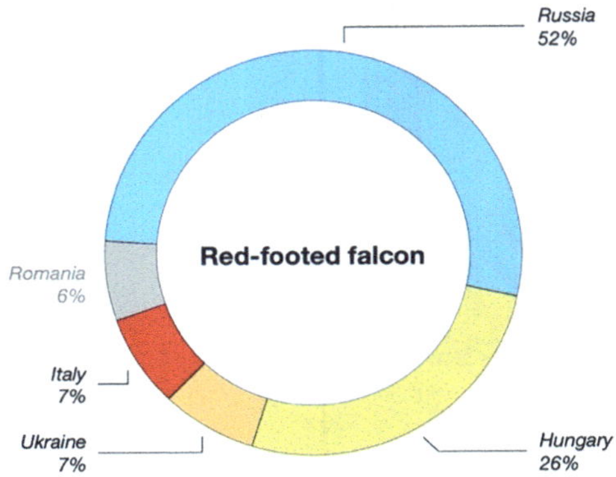

Fig. 6.9 Male, female, and *juvenile* of the red-footed falcon. (Naumann et al. 1899; Source: BHL https://www.biodiversitylibrary.org). Percentage of red-footed falcon specimens from the Late Modern Period conserved in major Natural History Museums. (Source: GBIF, http://www.gbif.org/)

6.4.4 Falco eleonorae

Eleonora's falcon was found in Sardinia, Greece, Syria, and parts of Africa (Bree 1859). According to Prince Bonaparte, it nested in July and August in cavities sheltered by bushes on rocky cliffs near the sea (Bree 1859). Hartert (1912) reported that this falcon inhabits rocky islands in the Mediterranean and along the western coast of Morocco, displaying a strong preference for specific habitats. In Morocco, it occurs in large numbers on the cliffs near Mogador and in the Haha province. Hartert (1912) also documented its presence on Lanzarote, where it certainly breeds, as well as on Fuerteventura. He mentions that some individuals are darker or more slate-gray, with intermediate forms between them and the reddish-brown forms below (Fig. 6.10). The black plumage is neither a difference of age nor sex, but it is rarer than the reddish-brown, "lark-falcon-like" variety (Hartert 1912). For him, it is clear that young blackish birds remain blackish throughout their lives, and the juvenile plumage transitions into the adult plumage solely through moulting (Hartert 1912).

6.4.5 Falco concolor

Although confusion with Eleonora's falcon and the gray kestrel persists, this previously unrecognized taxon is clearly identified during this period. Indeed, Bree (1859) stated that H. Schlegel, in his "*Revue*" in 1844, and other writers since then have suggested that C. J. Temminck confused specimens of the Eleonora falcon

Fig. 6.10 Eleonora's falcon in both morphs. (Naumann et al. 1899; Source: BHL https://www.biodiversitylibrary.org). Percentage of Eleonora's falcon specimens from the Late Modern Period conserved in major Natural History Museums. (Source: GBIF, http://www.gbif.org/)

with those of *F. concolor* (Bree 1859). Excluding all specimens classified as *F. concolor* from areas outside the species' range, very few specimens are from breeding countries (see Sect. 6.2.1; Gould 1837). In fact, most specimens originated from migrants (i.e., Ethiopia) but primarily from the Madagascar wintering grounds. Doubts still remained as to whether *F. concolor* was a European species or not, but there can be no doubt that the species are completely distinct (Bree 1859).

6.4.6 Falco columbarius

Nesting among the heather or on the ground, the merlin is regarded as a visitor to England in October and a breeding species in Scotland (Fig. 6.11; Fleming 1828; Gould 1837). During the nineteenth century, plover-netters operating in the Lincolnshire Wash and Humber frequently captured peregrines and merlins, and some of these birds were subsequently sold to falconers (Lorand and Atkin 1989). Hartert (1912) reported that this species breeds in northern regions, including northern Scandinavia, northern Russia, and Western Siberia, extending southward to central Russia. Additionally, it is known to breed on various islands and in parts of Western Europe, such as Iceland, the Faroe Islands, Scotland, Ireland, and Wales. In England, breeding sites include the Pennine Mountains, Derbyshire, and North Yorkshire, with Exmoor also being alleged as a breeding location. Recognized as a

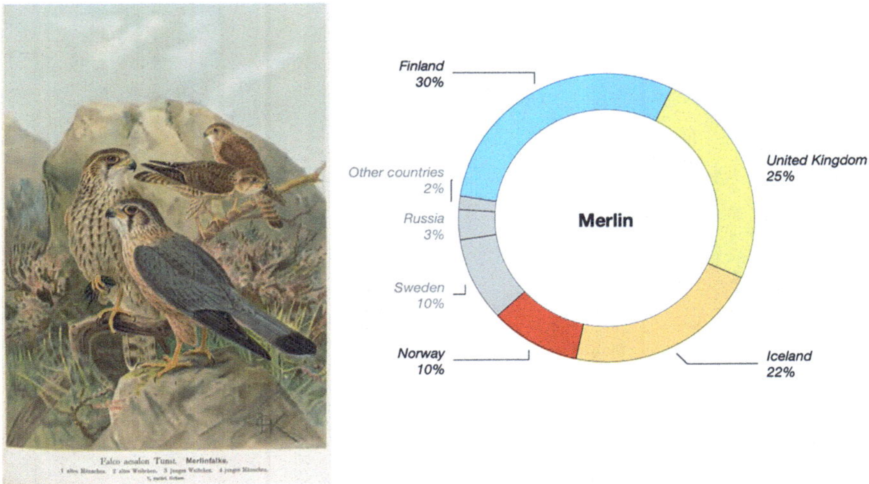

Fig. 6.11 Male, female, and *juvenile* merlin. (Naumann et al. 1899; Source: BHL https://www.biodiversitylibrary.org). Percentage of merlin specimens from the Late Modern Period conserved in major Natural History Museums. (Source: GBIF, http://www.gbif.org/)

migratory bird, it passes through Helgoland, Germany, and other parts of Central Europe during autumn and spring, wintering primarily in southern Europe. Smaller numbers are found in North Africa (Atlas Mountains) and larger concentrations in Egypt (Hartert 1912). It also winters in Great Britain, Ireland, the Faroe Islands, and Germany, though only in small numbers, likely confined to milder winter regions (Hartert 1912). It primarily lives in open terrain, bogs, tundras, coastal rocks, and similar habitats. It feeds mainly on songbirds, shorebirds, and similar species but also attacks larger birds and sometimes catches large moths. Although it has shorter wings than the lark falcon and is not as agile in the air, it is still fast and powerful. The eggs are laid in small depressions in moss or heather, either unlined or lightly lined, or on rocky ledges and in holes. It also sometimes takes over nests of crows and other birds (Hartert 1912).

6.4.7 Falco subbuteo

Gould regarded the Eurasian hobby as one of the most common smaller birds of prey in Europe, known for its wide range and migratory habits (Fig. 6.12; Gould 1837). As a tree-nester, it prefers wooded areas and occasionally reuses crow nests (Fleming 1828; Gould 1837). Ornithologists have noted its presence across Europe, from Scandinavia and northern Russia to the Mediterranean and some islands, reaching at least as far east as Western Siberia and the Altai. Hartert (1912) reported it as rare in England, sporadic in Scotland, and absent in Ireland. It overwinters in

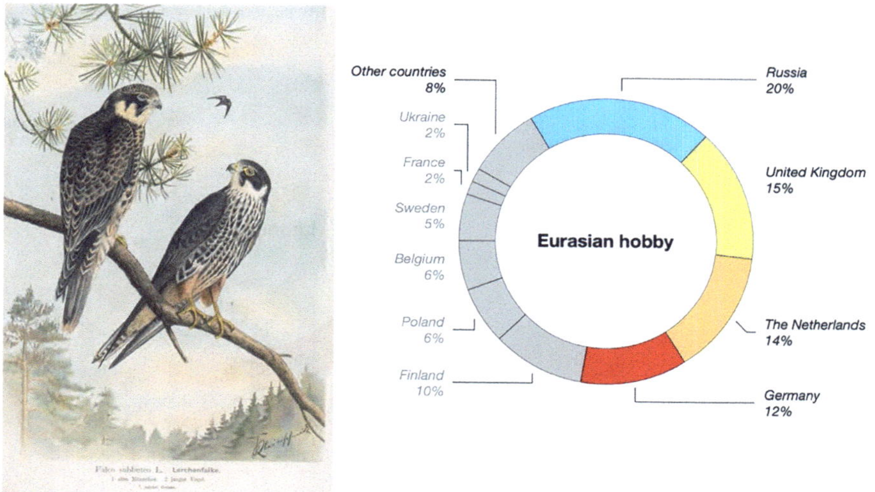

Fig. 6.12 Adult and *juvenile* European hobby. (Naumann et al. 1899; Source: BHL https://www.biodiversitylibrary.org). Percentage of Eurasian hobby specimens from the Late Modern Period conserved in major Natural History Museums. (Source: GBIF, http://www.gbif.org/)

tropical Africa, as far south as the Cape Colony (Hartert 1912). During the breeding season, it prefers light forests, forest edges, woodlands, and large parks. Its diet includes birds and insects like dragonflies, grasshoppers, and large beetles. This falcon mainly preys on larks and occasionally swallows (Hartert 1912). Swifts are rarely caught, usually when they are young or sick. For nesting, it uses the nests of crows, ravens, magpies, and other birds, often in trees, especially pines.

6.4.8 Falco biarmicus

It is considered a rare bird native to Eastern Europe and nearby areas in Asia and Africa (Gould 1837). In Western Europe, it is a vagrant and is rarely seen in France and the Netherlands (Gould 1837). Unlike Fleming (1828), who identifies it as a nesting species in Ireland and notes captures in Lincolnshire, Gould (1837) asserted that it never visits Great Britain. Furthermore, he viewed the lanner falcon as a completely distinct species rather than merely a variant of the peregrine falcon's juvenile plumage (Gould 1837). The modern history of the taxonomy and systematics of large falcons of the *Hierofalcon* subgenus is closely tied to the Rijksmuseum van Natuurlijke Historie of Leiden (RMNH) and its first two directors, Coenraad Jacob Temminck (1778–1858) and Hermann Schlegel (1804–1884; Hoek Ostende et al. 2008; Gassó Miracle 2008). The identification of the nominate species and the current subspecies is the outcome of their work, along with that of later German ornithologists. The numerous specimens, primarily from North Africa, have played a

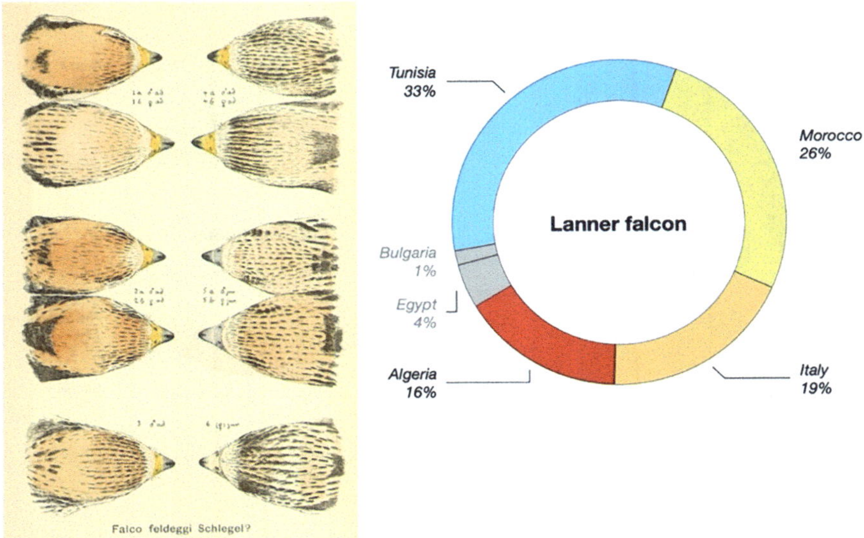

Fig. 6.13 Study drawings of the different plumage patterns on the nape and head of lanner falcons captured in Tunisia. The question is raised as to whether they belong to the same subspecies found in Europe, *F. b. feldeggii*. (Erlanger 1898; Source: BHL https://www.biodiversitylibrary.org). Percentage of lanner falcon specimens from the Late Modern Period conserved in major Natural History Museums. (Source: GBIF, http://www.gbif.org/)

key role in the accurate identification of those from Europe as well (Fig. 6.13). Accordingly, the reported sightings in Central and Northern Europe will be dismissed in favor of a more restricted southern range (Bree 1859). Expeditions in North Africa helped clarify the distribution of local subspecies. The species was considered present in the Atlas countries, more common in the south, less so in the northern Atlas, and also breeding in northern Tunisia, near Tangier, and possibly in southern Spain (Hartert 1912).

6.4.9 Falco cherrug

This species, more often linked to temperate than cold climates, was thought to inhabit the Asian steppes and parts of Europe, such as Hungary and southern Poland (Naumann et al. 1899). Bree (1859) notes that "*several individuals have been killed in Hungary, and young birds have been taken to Vienna in Austria.*" Woborzil (1846) described a nesting pair near Wetruschitz, now Čeradice in the Czech Republic, located in the northwestern region of the historical land of Bohemia. Fifty years after the discovery of this breeding population, it was considered extirpated in Bohemia, though it continued to inhabit suitable locations in the Austrian–Hungarian states (Naumann et al. 1899). A thorough historical analysis in Bulgaria indicates

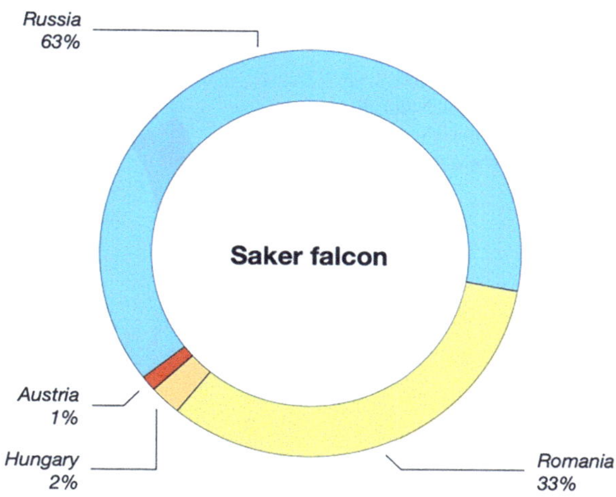

Fig. 6.14 Percentage of saker falcon specimens from the Late Modern Period conserved in major Natural History Museums. (Source: GBIF, http://www.gbif.org/)

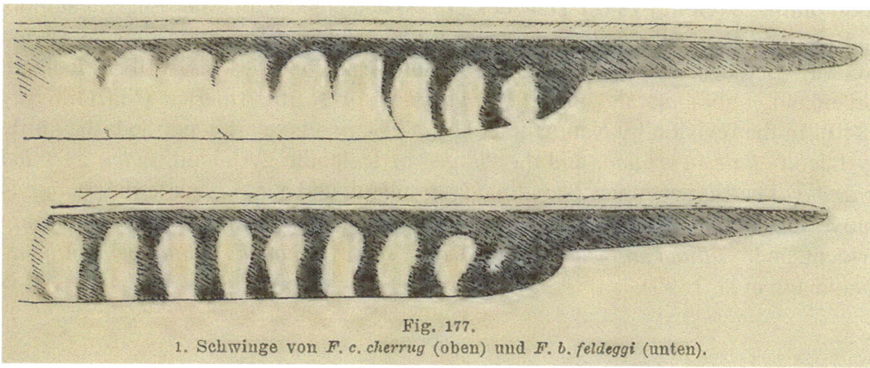

Fig. 6.15 Comparison of the shape and feather patterns of the primary feathers between the saker falcon and the lanner falcon (subspecies *feldeggii*; Hartert 1912)

that the species likely experienced a sharp decline in the nineteenth century, primarily driven by the intensification of agriculture and the expansion of human settlements (Ragyov et al. 2014). Misidentification between the saker falcon and lanner falcon continues to cause confusion in museum and private collections (Leonardi and Sutton 2020). Ornithologists use key morphological traits and plumage patterns to distinguish between the two species (Figs. 6.14 and 6.15)

6.4.10 Falco rusticolus

Gould (1837) describes it as a species from Northern Europe that lives on the west coast of Greenland and sometimes visits the Orkney and Shetland Islands. He also stated the various stages of transition from the brown *juvenile* plumage to the white

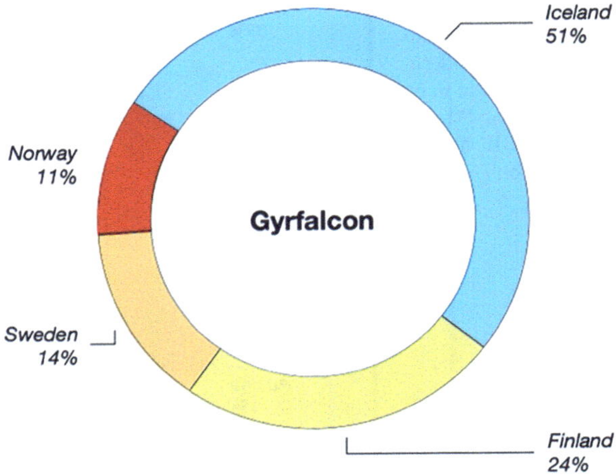

Fig. 6.16 Percentage of gyrfalcon specimens from the Late Modern Period conserved in major Natural History Museums. (Source: GBIF, http://www.gbif.org/)

adult plumage (Gould 1837). Hancock (1838) disagreed with other English authors and claimed there were different species of gyrfalcon. The wide range spanned Norway, Sweden, and several northern European countries, as well as Iceland, Greenland, Labrador, and the "*fur countries*" of North America (MacGillivray 1840). In the revision by Naumann et al. (1899), two taxa are identified: the "little gyrfalcon" *Falco gyrfalco*, and the "large" or Icelandic gyrfalcon, *Falco gyrfalco islandus*. The first occupies Scandinavia, Lapland, and Russia, while the second is native to Iceland (Fig. 6.16). However, significant similarities between them are evident, and "*without knowledge of its origin, some specimens cannot be identified*" (Naumann et al. 1899).

6.4.11 Falco peregrinus

With numerous specimens from the Americas and East Asia, ornithologists recognized the worldwide nature of the peregrine falcon. Differences in form and plumage patterns led to a debate about identifying potential species (Fig. 6.17). Reflecting a clear shift in the methodological approach from the past, Gould stated: "*However, if it could be proven that birds, such as the Magellanic and European Peregrines, could breed with each other and produce fertile offspring, we would then be constrained to regard them as simple but permanent varieties*" (Gould 1837). The transition from juvenile to adult plumage is recognized as the cause of the numerous supposed species and the various names assigned to the peregrine falcon (Gould 1837). Early observations on the spacing of nests include Eyton's (1838) note that he never saw peregrine nests in Wales located closer than 3 km apart. Hartert (1912) considered *F. p. peregrinus* not to be a migratory bird, but rather a mostly resident

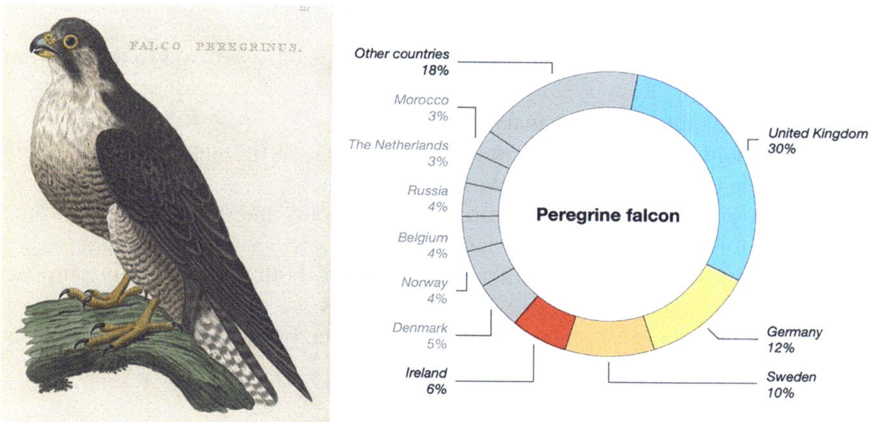

Fig. 6.17 An adult peregrine falcon depicted in "*Nederlandsche Vogelen*" (Dutch Birds), the first comprehensive avifauna of the Netherlands, which temporarily included Belgium from 1815 to 1830. This work by Cornelius Nozeman and Christiaan Sepp was completed in 1829. (Source: BHL https://www.biodiversitylibrary.org). Percentage of peregrine falcon specimens from the Late Modern Period conserved in major Natural History Museums. (Source: GBIF, http://www.gbif.org/)

and partial migrant species, which, except during the breeding season, roams more widely. It occasionally visits the Faroe Islands and, in winter, Sardinia, where *F. p. brookei* nests. Larger birds are definitely preferred, such as pigeons, partridges, grouse, lapwings, crows, shorebirds, and ducks, and other waterfowl, seabirds, kestrels, and small birds are also not ignored. *F. p. calidus* was identified as the northern peregrine falcon form that migrates far south in the autumn (Hartert 1912).

6.5 Conclusions

Over the 150 years of the Late Modern Period, two major developments in falcon research took place. In the nineteenth century, the large collection of specimens and the ability to preserve them led to a better understanding of taxonomy and systematics, as well as the distribution and abundance of species. On the other hand, many new taxa were created, often based on meristic differences in small sample sizes of specimens. A key development was the production of ornithological manuals, which shifted from being collections of individual experiences to organized, scientific compilations of information. In the 1930s, ethology emerged as a field focused on the study of animal behavior, marking the second shift in research. Researchers aimed to turn incidental observations into a more systematic understanding by using quantitative analysis and mathematical tools. This new direction will be fully realized in the contemporary period (Chap. 7).

Several key points can be highlighted:

- The taxonomy and systematics of the *Falco* genus have evolved to their current classification.
- Taxidermy enables in-hand analysis of falcons.
- The vast field collection of specimens is used to assess the status and distribution of individual species across the Western Palearctic.
- Ornithological manuals now feature modern layouts and organization, setting a standard for other contemporary handbooks.
- Falcons continue to face significant persecution, both for taxidermy purposes and because they are still seen as harmful species.
- Growing attention is being given to the need for their protection and conservation.

References

Allsen TT (2006) The royal hunt in Eurasian history. University of Pennsylvania Press, Philadelphia

Anonymous (1893) Travaux inédits—Nouvelle espèce européenne du genre Faucon. Revue Zool Soc Cuvierenne 2:105

Arrigoni degli Oddi E (1929) Ornitologia italiana. Hoepli, Milan

Barr W (2012) Gyrfalcons to Germany: Herdemerten's expedition to west Greenland, 1938. Polar Record 48:113–122. https://doi.org/10.1017/S0032247410000598

Belon P (1555) Histoire de la Nature des Oyseaux. Gillaume Cavellat, Paris

Braaksma S, Knippenberg WHT, Langenhoff V (1959) Enige broed-vogels in Noord- Brabant. Limosa 32:206–212

Bree CR (1859) A history of the birds of Europe not observed in the British Isles. Groombridge and Sons, London

Brehm CL (1827) Ornis. A Schmid, Jena

Cochrane JH (1864) Note on the nesting of the Lanner Falcon (*Falco lanarius*, Schlegel). Ibis 6:183–184. https://doi.org/10.1111/j.1474-919X.1864.tb07861.x

Cramp S, Simmons KEL (1980) The birds of the Western Palearctic, vol 2. Oxford University Press, Oxford

Cuvier G (1829) The Animal Kingdom, class Aves, vol 1. Whittaker, London

Dresser HE (1881) The birds of Europe including all the species inhabiting the Western Palearctic region, vol VI. Printed privately, London

Elzen R (2010) Type specimens in the bird collections of the Zoologisches Forschungsmuseum Alexander Koenig, Bonn. Bonn Zool Bull 59:29–77

Erlanger CF (1898) Beiträge zur Avifauna Tunesiens. J Ornithol 46:377–497. https://doi.org/10.1007/BF02208467

Eyton TC (1838) XXXII.—An attempt to ascertain the fauna of Shropshire and North Wales. Ann Mag Nat Hist 1:285–293. https://doi.org/10.1080/00222933809512293

Farber PL (1977) The development of taxidermy and the history of ornithology. Isis 68:550–565. https://doi.org/10.1086/351874

Fleming J (1828) A history of British animals. Bell & Bradfute, Edinburgh

Gassó Miracle ME (2008) The significance of Temminck's work on biogeography: early nineteenth century natural history in Leiden, The Netherlands. J Hist Biol 41:667–716. https://doi.org/10.1007/s10739-008-9158-8

Gené CG (1840) Descrizione di un nuovo falcone di Sardegna (*Falco Elenorae*). Tipografia reale, Turin

Gilbert HA (1927) Notes on the nesting habits of the peregrine falcon. Brit Birds 21:26–30
Glutz Von Blotzheim N, Bauer KM, Bezzel E (1971) Handbuch der Vögel Mittleuropas, vol 4. Akademische Verlegsgesellschaft, Frankfurt am Main
Gould J (1837) The birds of Europe. Vol 1 Raptores. Published by the Author, London
Gray R (1871) Birds of the West of Scotland. Thomas Murray and Sons, Glasgow
Hancock J (1838) XXVII.—Remarks on the Greenland and Iceland falcons, showing that they are distinct species. Ann Mag Nat Hist 2:241–250. https://doi.org/10.1080/00222933809496669
Hartert E (1912) (published in 1922) Die Vögel der palÄarktischen Fauna systematische Übersicht der in Europa, Nord-Asien und der Mittelmeerregion vorkommenden Vögel, vol II. FriedlÄnder, Berlin
Hartert E (1915) Notes on falcons. Nov Zool 22:167–185
Hoek Ostende LW, Dekker RWRJ, Keijl GO (2008) Type-specimens of birds in the National Museum of Natural History, Leiden. NNM Tech Bull 1:1–248
Hunt J (1818) British ornithology, vol I. Published by the author, Norwich
Ingram GCS, Salmon HM (1929) Notes on the nesting habits of the peregrine falcon (2). Brit Birds 22:198–202
Kleinschmidt O (1901) Der Formenkreis Hierofalcon die Stellung des ungarischen Würgfalken in demselben. Aquila 8:1–48
Koenig A (1892) Zweiter Beitrag zur Avifauna von Tunis. J Ornithol 40:329–416. https://doi.org/10.1007/BF02252121
Krueper T (1862) Ornithologische Notizen über Griechenland. J Ornithol 10:360–379; 435–448. https://doi.org/10.1007/BF02014951
Krueper T (1864) Beitrag zur Naturgeschichte des Eleonorenfalken, *Falco eleonorae* Gene. J Ornithol 12:1–23. https://doi.org/10.1007/BF02004572
Laurop CP, Fischer VF (1818) Sylvan, ein Jahrbuch fur Forstmänner, Jäger und Jagdfreunde. Krieger, Marburg and Cassel
Leonardi G (2001) Lanner falcon. BWP Update 3:161–178
Leonardi G, Sutton LJ (2020) *Falco lanarius*: the contorted history of a poetical archetype and a taxonomic puzzle. Archaeol Anthropol Sci 12:22. https://doi.org/10.1007/s12520-019-00995-1
Lorand S, Atkin K (1989) The Birds of Lincolnshire and South Humberside. Leading Edge, Burtersett
Lorenz K (1935) Der Kumpan in der Umwelt des Vogels. J Ornithol 83:137–289. https://doi.org/10.1007/BF01905355
MacGillivray W (1840) Birds of Great Britain and Ireland. Scott, Webster & Geary, London
Malan SC (1848) A systematic catalogue of the eggs of British Birds. John van Voorst, London
Marini MÂ, Hall L, Bates J, Steinheimer FD, McGowan R, Silveira LF, Lijtmaer DA, Tubaro PL, Córdoba-Córdoba S, Gamauf A, Greeney HF (2020) The five million bird eggs in the world's museum collections are an invaluable and underused resource. Auk 137:ukaa036. https://doi.org/10.1093/auk/ukaa036
Morris PA (1993) An historical review of bird taxidermy in Britain. Arch Nat Hist 20:241–255. https://doi.org/10.3366/anh.1993.20.2.241
Mykrä S, Vuorisalo T, Pohja-Mykrä M (2005) A history of organised persecution and conservation of wildlife: species categorizations in Finnish legislation from medieval times to 1923. Oryx 39:275–283. https://doi.org/10.1017/S0030605305000797
Naumann JA (1820) Naturgeschichte der Vögel Deutschlands. G. Fleischer, Leipzig
Naumann JA, Naumann JF, Hennicke CR (1899) Naturgeschichte der Vögel Mitteleuropas, vol 5. Köhler, Gera-Untermhaus
Newton I (1979) Effects of human persecution of European raptors. Rap Res 13:65–78
Nice M (1933) Some ornithological experiences in Europe. Bird-Banding 4:147–154
Orlando C (1957) Contributo allo studio del Lanario (*Falco biarmicus*) in Italia. Riv Ital Orn 27:147–153
Pearsall WH (1950) Mountains and Moorlands. Collins, London
Peters JL (1931) Check-list of birds of the world, vol I. Harvard University Press, Cambridge

Ragyov D, Biserkov V, Gradev G, Ivanov I, Stoynov E, Stoyanov G, Domuschiev D, Dixon A (2014) Past and present status of the saker falcon, *Falco cherrug* (*Aves*: *Falconidae*) in Bulgaria. Acta Zool Bulg 66:299–308

Rüppel E (1845) Vögel Nord-Ost-Afrika's. Frankfurt

Schlegel H (1844) Revue critique des oiseaux d'Europe. Roret, Paris

Schlegel H, Verster van Wulverhost AH (1845–1853) Traité de Fauconnerie. Leiden

Scholze W (1933) Ein Beitrag zur Fortpflanzungsbiologie des Baumfalken (Falco s. subbuteo L.). J Ornithol 81:377–388. https://doi.org/10.1007/BF01905456

Sharpe RB (1873) On the Peregrine Falcon from Sardinia. Ann Mag Nat Hist 11:20–21. https://doi.org/10.1080/00222937308696757

Temminck CJ (1825) Faucon concolore. *Falco concolor* Temm. Vol 1, Planche 330. In: Temminck CJ, Laugier M (1820–1839). Nouveau recueil de planches coloriées d'oiseaux. FG Levrault, Paris

Tinbergen N (1932) Ethologische Beobachtungen am Baumfalken Falco subbuteo. J Ornithol 80:40–50. https://doi.org/10.1007/BF01995021

Tinbergen L (1940) Beobachtungen tiber die Arbeitsteilung des Turmfalken (*Falco tinnunculus* L.) während der Fortpflanzungszeit. Ardea 39:63–98

Vaurie R (1961) *Falco eleonorae*. Ibis 103A:114–128. https://doi.org/10.1111/j.1474-919X.1961.tb02424.x

Verreaux J, des Murs O (1862) Étude sur l'identification del l'*Hypotriorchis concolor* et l'*H. eleonorae*. Rev Mag Zool May 1862

Walter H (1979) Eleonora's Falcon: adaptations to prey and habitat in a social raptor. Chicago University Press, Chicago

Warthausen RK (1860) On the nidification of certain birds in north-eastern Africa. Ibis 2:122–130. https://doi.org/10.1111/j.1474-919X.1860.tb06358.x

Woborzil JWE (1846) Beitrag zur Naturgeschichte der Raroh, *Falco lanarius*, Pallas. Rhea 1:39–43

Yalden DW, Albarella U (2009) The history of British birds. Oxford University Press, Oxford

Chapter 7
Contemporary History

7.1 The Systematic Persecution

In 1965, Prof. Karel Hendrik Voous (1920–2002) reported that, over the past 150 years, raptor populations in Europe had likely declined to just 1% of their original numbers (Bijleveld 1974). However, significant gaps in knowledge about the distribution and status of many species were also acknowledged (Bijleveld 1974). "*Birds of Prey in Europe*" is one of the first in-depth books on this topic, written by Maarten Bijleveld, a former student of Prof. Voous (Bijleveld 1974). This book essentially examines the persecution of species classified as harmful before delving into the status of individual species. Historical records from past centuries reveal that many predator species were already in significant decline by the mid-twentieth century (see Chap. 6). The situation was further exacerbated by the widespread use of biocides and other pollutants, identified as primary drivers of the sharp declines in populations of both large and small falcons across Central and Northern Europe (Newton 1979; Ratcliffe 1993).

The book also highlights the most critical threats, including:

- Drastic changes in agricultural practices
- The extensive expansion of power lines
- The rapid growth of tourism

From 1975 to the present, these threats—along with numerous emerging challenges—have continued to be identified as the most significant factors impacting species, as reflected in conservation action plans developed for each case.

How was the status of falcons perceived 50 years ago? While a comparison with current data might seem minor, it provides valuable insights into the development of both knowledge and ideas. A broader perspective on this evolution is offered in the "*Preface*" of this book and the conclusions presented in Sect. 7.6.

G. Leonardi, *The History of Western Palearctic Falcons*,
https://doi.org/10.1007/978-3-031-86528-2_7

Among small falcons, the Eurasian hobby was reported to be declining across Europe, though to a much lesser extent than the peregrine falcon. This disparity was partly attributed to its diet, which included fewer seed-eating birds, potentially reducing its exposure to pesticides. The primary threat to merlins breeding in northern Europe appeared to have been persecution during migration and in their wintering areas, along with the risk of poisoning from biocides. Instances of red-footed falcons breeding west of their typical range in southeast Europe, likely due to east–west migration patterns, were documented in Scandinavia, Germany, and Austria. However, in this context, only population trends in their usual breeding areas were considered relevant.

The lesser kestrel remained numerous within its Mediterranean range but showed widespread signs of decline. Meanwhile, the common kestrel was the most abundant and widely distributed falcon in Europe. Despite this, its population had been significantly reduced throughout the continent due to hunting and biocide use.

The dramatic decline of the peregrine falcon in recent times, which brought the species to the brink of extinction, drew significant attention and ultimately highlighted the impact of pesticides on the natural environment. By then, only a few populations had survived, and the chances of the species avoiding extinction in continental Europe appeared extremely slim. During this period, the lanner falcon faced a similar threat of extinction in Europe, which represented the northernmost edge of its range. In contrast, the saker falcon, found in southern Europe, showed a slower rate of decline. This was likely due to its reliance on small mammals as a primary food source, making it less vulnerable to pesticide exposure. The limited presence of seed-eating birds in its diet further reduced its risk of biocide poisoning. Meanwhile, the gyrfalcon had already suffered a severe population decline in Scandinavia, beginning in the eighteenth and nineteenth centuries. Its present-day rarity in Denmark, where it was once a common visitor, serves as a stark reminder of this long-term decline.

A significant challenge in falcon conservation, both in the past and present, has been the ability to conduct a realistic assessment of viable populations. The data from the 1970s were sparse and fragmented, likely leading to an underestimation of population sizes, particularly for rare or elusive species (see also Sect. 7.6). At the same time, the ornithological community was becoming increasingly aware of the harmful effects of chemical compounds, which amplified concerns about the future of these species.

Contemporary history covers the period from 1945, including the second half of the twentieth century and the first quarter of the twenty-first century. In practice, it begins with the end of the Second World War in 1945, marking the rise of globalization and major technological advancements.

7.2 1945–1960

At the end of the Second World War, the perspective on species expanded significantly, embracing a more comprehensive and holistic ecological approach. The primary focus for raptors, particularly falcons, was on their essential role as predators. In the 1920s and 1930s, regurgitated pellets from raptors were first studied in laboratories, and during this time, they became recognized as indicators of the predators' dietary habit (Glading et al. 1943). David Lack (1910–1973), the renowned evolutionary biologist, conducted one of the first meta-analyses on the feeding habits of falcons in 1946, aiming to evaluate competition among predators. Lack, after examining extensive samples, concluded that closely related species do not compete for food within the same habitat (Lack 1946; Fig. 7.1). His most important observation regarding the absence of competition is that, in their natural state, predators are ultimately limited in population density by the availability of food (Lack 1946).

A further comprehensive analysis was conducted by the Swedish ornithologist Gustaf Rudebeck (1913–2005) on the feeding habits of predatory birds (Rudebeck 1951). His research, based on a substantial sample of hunting attempts in nature, shows that the merlin's success is only 14.3%, while the peregrine falcon achieves a success rate of 15.8% (Rudebeck 1951; Curry-Lindahl 1961a, b).

Lack's renowned paper on the breeding seasons of birds, following his previous work on predation, introduces a new comparative quantitative approach (Lack 1950). In particular, he expanded on and refined the concepts introduced by Baker (1938) regarding the "*ultimate*" factors that connect the breeding season to the availability of food for offspring, as well as the "*proximate*" factors that affect breeding conditions (Lack 1950). In Lapland, for example, the gyrfalcon, as Arctic species, lays its eggs about a month earlier than the other three species of *Falco*, including the peregrine falcon (Lack 1950). Likewise, the European hobby's breeding season is linked to late summer, when swallows, swifts, and dragonflies are abundant (Lack 1950). Eleonora's falcon is the furthest example, raising its young in September and October while taking advantage of the passage of small- and medium-sized migratory birds (Stresemann 1927–1934; Lack 1950).

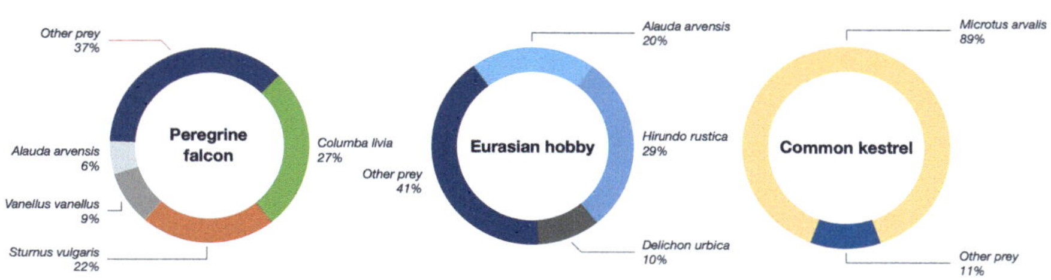

Fig. 7.1 Prey of different falcon species in Germany: peregrine falcon (5504 prey examined), Eurasian hobby (900 vertebrate prey examined), and common kestrel (30,000 prey examined; Uttendörfer 1939; Lack 1946)

During this period, the migratory patterns of falcons begin to emerge, with movements between the Palearctic and the Middle East (such as through the Suez Canal), as well as toward North-eastern Africa (Simmons 1951; Guichard 1957). Similarly, the ringing scheme, which started in the early 1900s, begins to reveal the migratory movements of falcons, including the common kestrel, between Northern and Central Europe (Price 1947; Landsborough Thomson 1958; Fig. 7.2). These individuals primarily winter in France, Belgium, and Spain, but some also spend the season in Sweden (Landsborough Thomson 1958; Fig. 7.2).

The analysis of data on the autumn passage of falcons at Falsterbo in Sweden during the war years shows minimal numerical variation (Fig. 7.3). According to Rudebeck (1950), despite significant persecution, the peregrine falcon remains relatively common as a passage migrant in Scania. Locally, the common kestrel increased during the war years (Price 1947). At Falsterbo, the low numbers of the common kestrel, initially caused by the cold winters during the Second World War, saw further decline in the 1950s, likely due to the harmful effects of pesticides, which persisted into the following decade (Kjellén and Roos 2000). The high numbers of merlins and peregrine falcons are similarly expected to decline sharply. For example, the Swedish population of peregrine falcons dropped from an estimated 350 pairs in 1950 to 35 pairs in 1965, reaching a low of just 15 breeding pairs by the mid-1970s (Kjellén and Roos 2000). In fact, a significant demographic bottleneck, such as that experienced by the peregrine falcon, occurred during the Second World

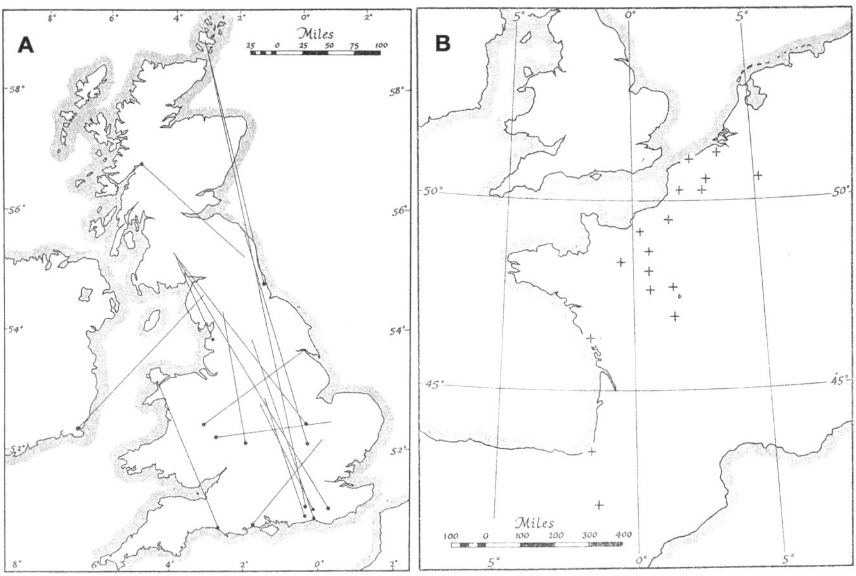

Fig. 7.2 (**a**) First-winter recovery of common kestrels within Great Britain and Ireland. (**b**) Recoveries of first-winter individuals ringed as nestlings in Great Britain and the Isle of Man. (Thomson 1958; Source: https://www.biodiversitylibrary.org/bibliography/105650)

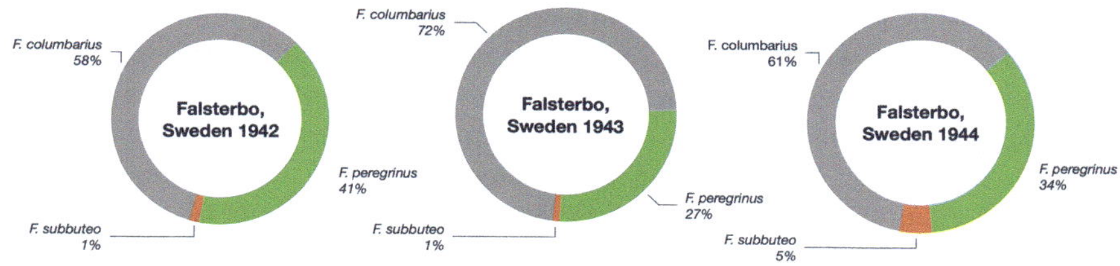

Fig. 7.3 Autumn migration of falcons across the Falsterbo Peninsula, located in the southwestern corner of Sweden, during the Second World War (1942–1944; Rudebeck 1950)

War, extending the impact of persecutions that began in the eighteenth century (Jacobsen et al. 2008).

The most important and seminal study of this period comes from the other side of the ocean: Tom Cade's monograph on the peregrine and gyrfalcon in Alaska (Cade 1960). It exemplifies a comprehensive and systematic approach to understanding the reproductive biology of these species. Overall, the peregrine falcon was recognized as a species with a wide ecological niche (Svärdson 1949). Ornithologists noted that its ability to nest in various habitats (rock cliffs, trees, and bare soil) and its distribution across northern Europe have allowed it to occupy various territories (Thomasson 1947; Svärdson 1949). Dementiev and Gortchakovskaya (1945) argued that gyrfalcon populations, naturally limited to a few nesting pairs, should not be seen as harmful. On the other hand, in 1941, they reported that poachers likely disturbed the gyrfalcons, preventing them from raising their young and highlighting the negative effects of such activities. Another study conducted in Iceland suggests a connection between the gyrfalcon's early breeding season and the breeding seasons of ptarmigan and ducks (Wayre and Jolly 1958). Among small falcons, Lajos Horváth's study on the red-footed falcon in Hungary's Ohat Forest stands out as a pioneering example of a modern approach to small falcon research (Horváth 1955). This comprehensive work delves into key aspects of the species' biology, including detailed observations on breeding performance, nesting distances, and distribution patterns. The study follows the falcons' life cycle across the breeding season, from their initial arrival and mate-pairing to their departure during the post-fledgling period (Horváth 1955).

In the 1950s, falcon populations faced a significant threat of extinction. Carl Demandt's detailed study of the peregrine falcon in Germany brought this alarming trend to light (Demandt 1953). Through his review of ornithological literature, he identified the peregrine falcon as the most closely studied native bird of prey. He attributed this focus to two primary factors: the species' elegance, which fascinates field ornithologists, and its growing rarity. Despite existing protective laws, persistent persecution had severely diminished its numbers, drawing increased attention to its conservation needs. In Westphalia, observations since the end of the war have shown that out of roughly 10 breeding pairs, only 1–2 have successfully raised fledglings (Demandt 1953). Additionally, nests often contain only two eggs instead of the usual clutch size of 3–4, with one egg frequently found to be unfertilized

(Demandt 1953). From 1949 to 1956 in Great Britain, a significant number of peregrine falcon eyries were found to contain broken eggs (Ratcliffe, 1958). After considering all possible explanations, the true causes, however, remained unclear (Ratcliffe, 1958). During the 1950s, organochlorine pesticides were viewed as much safer than the arsenicals in use at the time, which frequently caused animal toxicity (Smits and Naidoo 2018). Nevertheless, as noted by Wegner et al. (2005), the recorded decline, which occurred prior to the use of DDT, along with the lack of offspring, likely indicated an aging of the breeding population.

7.3 1961–1970

7.3.1 Systematics

The classification of species, as outlined in Table 6.2, provides the foundation for further discussion on some subspecies, which were largely defined in the previous century. Among the peregrine falcons, three Palearctic subspecies are reaffirmed: *F. p. peregrinus*, *F. p. calidus*, and *F. p. brookei* (Vaurie 1961a). Regarding the so-called barbary falcon, it was initially considered a distinct species comprising two subspecies: *F. pelegrinoides pelegrinoides* and *F. p. babylonicus* (Vaurie 1961a). However, subsequent genetic studies suggest that the barbary falcon is more accurately classified as a subspecies of the peregrine falcon rather than a separate species (Wink and Seibold 1996; White et al. 2013; Table 6.2). The saker falcon's nominal subspecies, *Falco c. cherrug*, inhabits the European region of the western Palearctic (Vaurie 1961b). In contrast, the supposedly distinct species *F. altaicus* is more likely an ecotype or a variant of the eastern saker falcon populations (Vaurie 1961b; Zinevich et al. 2023). The debate over the correct identification of potential gyrfalcon subspecies (ranging from four to seven on average), which began in the previous century, continued into the mid-twentieth century. Ultimately, based on the literature and measurements from a large series of individuals, Vaurie (1961b) concluded that "*it seems wiser not to recognize any subspecies.*" The three subspecies of the lanner falcon, established in the nineteenth and early twentieth centuries, remain valid (Leonardi 2015). However, the two subspecies from North Africa and the Middle East (*F. b. erlangeri* and *F. b. tanypterus*) should be regarded as identical, given their similar plumage patterns and the genetic admixture observed between them (Clark 1999; Attili et al. 2023).

7.3.2 Behavioral and Population Ecology

Ornithologists are increasingly focusing on lesser-known bird species that inhabit remote and underexplored areas, such as small islands and hyper-arid regions. Their efforts include detailed reviews of existing knowledge, such as studies on Eleonora's falcon, as well as significant discoveries like the first confirmed photographs of the sooty falcon nesting in the Libyan desert (Booth 1961; Vaughan 1961). Eleonora's falcon is a unique species with distinctive behavioral traits, including being an extremely localized colonial breeder and feeding exclusively on small migratory birds (Spina and Leonardi 2007). Due to these characteristics, the species is an ideal candidate for collecting extensive data from a large population, while requiring only a reasonable field effort. Stresemann (1968) was among the first to estimate the global population of this falcon and assess its potential impact on the abundance of migratory birds. The comprehensive study conducted by H. Walter in two colonies in Crete and along the Atlantic coast of Morocco provides a significant early example of using field data to develop a thorough understanding of the species' reproductive biology (Walter 1968). The paper also includes numerous photographs and presents the first distribution map of the species within the Mediterranean basin (Walter 1968). A. J. Cave's study on the nesting of common kestrels in the Netherlands marked another significant shift in falcon research (Cavé 1968). By using nest boxes that the common kestrels readily adopted for breeding, Cavé was able to gather a wealth of data with reasonable effort (Cavé 1968). To understand the effects of prey abundance, he conducted a thorough analysis of the pellets, along with an evaluation of breeding performance and the condition of the females, particularly the development of their oocytes (Cavé 1968). This approach facilitated the exploration of various interconnected research questions and led to a surge in complex studies that began in the 1980s and 1990s.

Research on migratory routes is expanding to include several significant migration bottlenecks, such as the Bosphorus (Porter and Willis 1968), Eilat (Safriel 1968), Malta (De Lucca 1969), and Lebanon (Nielsen and Christensen 1970). Following a decade of research on the partial migration of British common kestrels from the United Kingdom to Central Europe (see Fig. 7.2), researchers are beginning to uncover key insights into the composition of the migrating populations and the underlying causes of this phenomenon. Young birds were found to be more migratory than adults, with their movements linked to areas experiencing fluctuations in vole populations (Snow 1968). Furthermore, the annual mortality rate of common kestrels was estimated at approximately 60% during their first year of life, decreasing to around 34% from the third year onward (Snow 1968).

The recognition of the sharp decline in falcon species led to both a need for understanding the status and size of national populations (cf. Ratcliffe 1962; Curry-Lindahl 1970; see also Box 7.1) and the emergence of local initiatives (Prestt 1965; Griffiths 1967; Gilbertson 1969). For example, a questionnaire featuring a data validation system was distributed to experts and associations (Prestt 1965). Another initiative was local surveys, such as the common kestrel breeding population survey in Leicestershire in 1963 and 1964, which was primarily based on volunteer efforts (Griffiths 1967). Data were organized using a square grid map and a standardized collection protocol. Initially conducted at the county or provincial level, this practice expanded to entire regions in the 1980s and 1990s, significantly improving the knowledge of falcon densities, despite remaining geographically fragmented (see Sect. 7.4). The census and monitoring of active peregrine falcon territories in Great Britain represent a significant field effort, marking the first of its kind (Ratcliffe 1962, 1963). This initiative has made it possible to conduct a meaningful comparison with data from the pre-Second World War period (Ratcliffe 1963). In addition, the study has allowed for an in-depth exploration of various aspects of reproductive biology, including breeding habitat requirements, breeding density, and territory size (Ratcliffe 1962).

Based on Rudebeck's comprehensive study (1951), the research focused on understanding the interplay between predators and prey, specifically examining how birds of prey influence prey populations and how prey species, in turn, affect predator populations. The feeding niches of falcons are typically distinct, with minimal overlap (Blondel 1967; Thiollay 1968). Additionally, predation imposes selective pressure that affects some species more than others (Blondel 1967; Thiollay 1968). On the other hand, prey numbers can fluctuate, significantly impacting falcons at both the individual and population levels (Frochot 1967). Moreover, feeding choice is recognized as influencing the intensity of reversed size dimorphism across different species (Frochet 1967).

7.3.3 Conservation

The recognition of the substantial negative effects of chemical compounds on falcon survival has brought renewed attention to long-standing issues. Human activities, such as deliberate extermination or unintentional disruption of breeding sites, have significantly contributed to the decline and local extinction of various bird species, including peregrines and gyrfalcons (Curry-Lindahl 1961b). In addition to these threats, falcons were historically targeted as harmful species in several European countries, including Italy, where such hunting practices were fully legal. For instance, Lovari (1974) documented 84 common kestrels shot over three hunting seasons, which were later sold in the taxidermy market. However, while the analysis of falcons' diets provides valuable insights, the debate remains clouded by persistent biases about whether these predators pose a threat to game populations (Bergman 1961; Curry-Lindahl 1961a). During this period, increasing emphasis is placed on the significant impact of trafficking falcons and their eggs—whether for legal or illegal falconry markets—on species that are already declining or naturally rare and localized (Bijleveld 1974). The trade is highly profitable, often disregarding laws and the increasing rarity of birds of prey, which only drives their prices higher. As a result, organized expeditions are dispatched to capture gyrfalcons from the North, saker falcons from the East, lanner falcons from the South, and peregrine falcons from any available location (Géroudet 1969). According to Géroudet (1969), in 1968, German "falconers" returned from Sicily and Italy with nearly 40 lanner and peregrine falcons, all taken from nests, and sold them for prices ranging from 500 to 1000 marks each.

The conservation legislation for these species varied widely across European countries. For example, Sweden, as early as the 1950s, included all species of the genus *Falco* on its protected list, providing them with full protection (Curry-Lindahl 1961a).

Box 7.1: Environmental Pollutants and Their Relationship with Falcon Populations

In the 1960s and 70s, falcon populations, especially large falcons, declined sharply across the Northern Hemisphere (Newton 1979; Ratcliffe 1993). This was due to persecution and the use of harmful insecticides like DDT, dieldrin, PCB, hexachloro benzene, and mercury (Newton 1979; Ratcliffe 1993; Smits and Naidoo 2018). In particular, the accumulation of these chemicals reduced the hatching success of peregrine falcons, resulting in a sharp population decline (a 90% reduction; Jacobsen et al. 2008). This falcon decreased to as few as 17 known pairs in 1976, with 15 remaining in northern Scandinavia and two remaining in southern Scandinavia (Lindberg 1977; Wille 1977; Lindberg et al. 1988). In 1961, most peregrine falcon nesting sites in southern England and Wales were deserted, with only 7% of pairs successfully rearing young across 156 sites (Cramp 1963). Since the 1970s, Norway has also monitored gyrfalcons for contaminants, especially in their eggs (Gjershaug et al. 2008). Among small falcons in Sweden, merlins have been found to contain pesticide residues, including aldrin and mercurial fungicides (Curry-Lindahl quoted in Cramp 1963). The widespread decline in Great Britain became apparent in the early 1960s, starting with more adaptable species such as the common kestrel (Cramp 1963). This decline was evident in southern and eastern England only when disentangled from national data (Fig. 7.4).

Early analyses showed that toxic chemicals caused some fatalities among individuals, but more significantly, that sublethal doses of these poisons could have devastating effects on fertility (Cramp 1963). Adult mortality was mainly caused by dieldrin and endrin (Bogan and Mitchell 1973; Nisbet 1988). Understanding the mechanism by which DDE, a degradation product of DDT, causes eggshell thinning took nearly 20 years (Ratcliffe 1967,1970; Peakall 1975). On average, the shell thickness index decreased by 23.3% compared to values recorded before 1947 (Nygård et al. 2019). The rate of egg breakage among peregrine falcon breeding pairs increased dramatically, rising from 2.8% between 1904 and 1950 to 27% during the 1951–1966 period (Ratcliffe 1967).

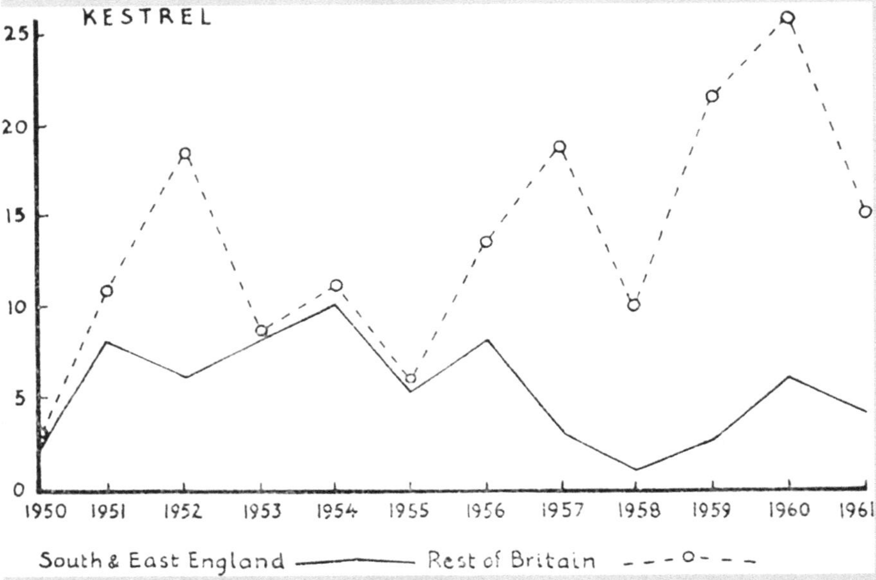

Fig. 7.4 Trends in nesting records of common kestrels: a comparison between Great Britain as a whole and southern and eastern England (1950–1961; Cramp 1963; Source: https://www.biodiversitylibrary.org/bibliography/105650)

The ban on DDT and restrictions on mercury use have helped peregrine falcons recover from near extinction in the mid-1970s to stable populations today (Smits and Naidoo 2018). DDT and dieldrin were banned in Norway in 1970, followed by PCB in 1980 (Holt et al. 1979). In the UK, dieldrin was specifically banned as a seed-dressing agent in 1975 (Walker and Newton 1999). Despite remaining relatively thin throughout the 1970s, 1980s, and 1990s, eggshells have since increased to nearly normal levels in recent decades (Wegner et al. 2005; Nygård et al. 2019). Because DDT is being phased out more slowly in the peregrines' wintering grounds in Latin America, pre-DDT shell thickness in Greenland is expected to be reached around 2034 (Falk et al. 2018).

7.4 1971–1980

This decade marks a pivotal turning point in the contemporary period, separating the current phase from the previous one. It lays the groundwork for a new, comprehensive approach to species and, more importantly, focuses on the sustainable management of populations.

7.4.1 Handbooks and Monographs

The development and recognition of this new approach are reflected in key publications on falcons, particularly two major ornithological manuals: "*Handbuch der Vögel Mitteleuropas*" and "*The Birds of the Western Palearctic*." The first was published at the start of the decade, while the second appeared at its close (Glutz et al. 1971; Cramp and Simmons 1980). Both works compile knowledge gained over the last 20–25 years, alongside a wealth of scattered sources. They were created with the contributions of numerous collaborators and continue to serve as essential references. Notably, "*The Birds of the Western Palearctic*" includes the first detailed distribution maps for all falcon species (Cramp and Simmons 1980). These works mark the culmination and the end of similar efforts from the nineteenth century (see also Sect. 6.2.2 and Box 6.1). The process of gathering information revealed substantial gaps in the knowledge of each species. It became clear that a generalist approach was not feasible; instead, species-specific studies were necessary. The first of these was the 1981 monograph on the peregrine falcon, the flagship species, completed by D. Ratcliffe (1993).

Indeed, the true precursor to the subsequent (and still ongoing) new approach in the study of falcons was Ian Newton's 1979 monograph "*Population Ecology of Raptors.*" It is also an extensive compilation of previous data, but instead of merely presenting the information, Newton uses it to develop a coherent understanding of the species, focusing on their adaptations to the environment (Newton 1979). Notably, the book's table of content does not list species but rather broad themes that link them, such as "*dispersion*," "*breeding strategies*," and "*mortality*."

7.4.2 Identification of Falcons

The growing conservation interest in raptors, particularly falcons, led to an increasing number of researchers and birdwatchers focusing on these species. This growing interest highlighted the need for a reliable tool to aid in proper field identification. As discussed in previous chapters, this has been a long-standing challenge, resulting in frequent misidentifications and inaccurate assessments of species' status across various countries. The work of Christensen and colleagues at the start of the 1970s marked a pioneering effort in creating an illustrated guide specifically for identifying these predators in flight (Christensen et al. 1973a, b, c). Alongside the illustrations, and benefiting from improved and more accessible technical equipment, the guide also includes numerous photographs (Christensen et al. 1973a, b, c). This project, later compiled into a volume, was made possible by observing a wide range of species during their migratory passage through the Middle East.

7.4.3 Behavioral and Population Ecology

Standardized data collection protocols have been developed, extending even to the national level, to monitor population trends in relation to threats and conservation efforts (Ratcliffe 1972). Additionally, reproductive parameters, such as brood size, are assessed to facilitate comparisons with the pre-1947 period (Ratcliffe 1972).

Increasing attention is being given to field research on rare species. In Italy, this includes efforts to define the northern range limit of the lanner falcon and study its reproductive biology in this boundary region (Chiavetta and Bonora 1973; Bonora and Chiavetta 1975).

Research has been started to explore the differences between gregarious and solitary breeding strategies (Horváth 1975). Studies on diet focus not just on listing prey but on understanding the predator–prey relationship. For example, the common kestrel is recognized as an opportunistic hunter rather than a specialist, typically feeding on a range of prey in smaller quantities, with its diet largely determined by the availability of alternative prey (Davis 1975).

The sharp decline in falcon species has created an urgent need for solutions, prompting the consideration of using captive-bred falcons for repopulation efforts. While wild falcons have been trained and held in captivity, sustained attempts to propagate these birds have only been made in this period (Cade et al. 1977). In 1942 and 1943, Waller (1962), a German falconer, succeeded in breeding a pair of peregrine falcons.

Launched in 1970, the Peregrine Fund's research program in the United States aimed to develop the necessary techniques for breeding falcons in captivity (Cade et al. 1977). Its goal was to establish a sufficiently large captive population, capable of providing enough birds to reintroduce breeding peregrine falcons into the wild (Cade et al. 1977). In Europe, breeding colonies of captive falcons, including the common kestrel, were established (Kirkwood 1980). The primary objective was to define reproducible techniques for housing, feeding, breeding, and managing the colonies (Kirkwood 1980). Having breeding pairs in captivity provided the opportunity to observe their behavior in detail and continuously throughout all stages of reproduction, a practice that was difficult to implement in the wild at the time (Wrege and Cade 1977). Furthermore, studies of captive individuals have revealed a significant gap in understanding the role of disease in the current status and ecology of falcons (Greenwood 1977).

7.4.4 Conservation

In the 1970s, 14 European countries granted full protection to all birds of prey, while 16 offered partial protection (covering certain species, regions, or seasons; Conder 1977). In contrast, one country, Malta, provided no protection at all (Conder 1977). Peregrine falcons remain among the most protected animals in the United

Kingdom, classified as a Schedule I species under the Wildlife & Countryside Act 1981 (Mak 2024). Reintroduction projects and the banning of harmful chemicals like DDT played a key role in boosting the number of wild peregrine falcon populations across Fennoscandia, Northern, and Central Europe, starting in the early 1980s (Jacobsen et al. 2008). In 1974, a reintroduction project began in southern Scandinavia to help the peregrine falcon population (Lindberg et al. 1988). Birds from Fennoscandia and Scotland were used in a captive breeding program to prevent extinction and protect genetic diversity (Lindberg 1977).

The potential impact of chemical compounds is also assessed in other species. Eleonora's falcon, a colonial species, is particularly suitable for such studies (refer to Sect. 7.3.2). Research involving egg analysis has shown that only a small proportion of the studied Eleonora's falcon population appears to be at risk from pesticide exposure (Clark and Peakall 1977). Outside the breeding season, Eleonora's falcon primarily feeds on insects, which likely results in adults arriving at breeding sites with minimal contaminant loads. However, the variability in contaminant levels among prey species means that the pesticide levels in the eggs are influenced by the prey consumed during a relatively brief time frame (Clark and Peakall 1977).

7.5 1980–Present

The 1980s and 1990s marked a pivotal period in falcon research, characterized by significant progress that built upon the new paths defined in the previous decade. During this time, much of the focus was on studying and monitoring species that were rare, geographically restricted, or less well-known, helping to bridge crucial gaps in understanding (Frumkin and Pinshow 1983; Negro et al. 1991; Hiraldo et al. 1996; Leonardi 2001). Controlled experiments, both in captivity and in the wild, enabled researchers to investigate a range of behavioral and physiological traits (Csermely 1993; Valkama and Korpimäki 1999). These studies also provided valuable insights into the immunology and toxicology of falcons, enriching our understanding of their biology (Daan et al. 1996; García-Fernández et al. 1997; Erdélyi et al. 1999). Over the years, falcon research evolved beyond its niche, becoming central to broader ecological studies (Donázar et al. 2016; McClure et al. 2018). By the 2000s, falcons were at the heart of research examining the effects of climate change and the spread of disease, serving as key indicators for global environmental shifts (Rodríguez and Bustamante 2003). Despite the considerable achievements, research has considerably slowed over the past two decades. Today, the focus has shifted toward the urgent task of conserving falcon species, as their survival is increasingly threatened by the extensive impact of human activities on their habitats (see Sect. 7.5.3).

7.5.1 Systematics

The *Falconidae* family has, after a long time, been recognized as distinct from the traditional, large group of diurnal birds of prey. This group includes eagles, hawks, the osprey, the secretary bird, and vultures, which belong to the *Accipitridae*, *Cathartidae*, *Sagittariidae*, and *Pandionidae* families (Hackett et al. 2008). Remarkably, falcons share a common ancestor not with these birds but with parrots (*Psittaciformes*), seriemas (*Cariamiformes*), and passerine birds (*Passeriformes*; Pyle 2013; Jarvis et al. 2014). Studies of both morphology and molecular biology indicate that the similarities observed among the so-called "raptors" are the result of convergent evolution (Hackett et al. 2008; McCormack et al. 2013; Yuri et al. 2013). As a result, key analogous features, such as strong beaks and sharp talons, evolved independently in different lineages. These traits likely arose either in *Australaves* and *Afroaves* or from an even more ancient landbird ancestor (*Telluraves*) with a raptorial lifestyle (McClure et al. 2019).

7.5.2 Reproductive Biology

Numerous studies have examined both proximate and ultimate factors affecting the falcon breeding performance, building on theoretical research from the 1950s (see Sect. 7.2). One key determinant of breeding success in falcons is the precise timing of reproduction. As day-length increases, several neuroendocrine and endocrine secretions are stimulated, triggering gonadal development in preparation for the breeding season (Meijer 1989; Meijer et al. 1989). In the Western Palearctic, longer photoperiods are associated with earlier laying dates in falcon populations (Meijer et al. 1989). Overall, hormonal patterns, reproductive behavior, and body mass indicate that female falcons have a broader reproductive window than males (Meijer 1989). Female falcons begin egg formation after accumulating body reserves, which are later depleted during the early nesting period. The timing of egg-laying is influenced by factors such as food availability in winter and spring, the amount of precipitation during these seasons, and spring temperatures (Cavé 1968; Potters 1998; Zellweger-Fischer et al. 2011). Moreover, feeding frequency and nestling growth rates are negatively correlated with daily rainfall (Kuusela and Solonen 1984). For instance, high rainfall in April has been shown to negatively impact the productivity of both peregrine and saker falcons (Chavko et al. 2014; Zuberogoitia et al. 2015).

Within a breeding population, researchers started to distinguish different components: (1) breeding individuals from previous years, (2) first-year individuals from the same area, and (3) floaters. Floaters played a fundamental role in population growth, especially in small-sized populations (Hunt 1998). Where to breed has a significant impact on fitness, so the occupancy of nest sites is nonrandom, reflecting a preference for certain territories that are likely of higher quality (Amato et al. 2014; Burke et al. 2015). However, when nest sites are abundant without a corresponding increase in food availability during the mating period, the number of breeding pairs does not increase (Wiklund and Larson 1994). Although core nesting locations represented only a small fraction of available sites, they accounted for the majority of nesting attempts, and the number of young fledged was greater than that at other sites (Gahbauer et al. 2015; Bruggeman et al. 2016). In lesser kestrels, 78% of all breeding attempts occurred in previously occupied nests, and these attempts were more successful than those in holes used only once (Negro and Hiraldo 1993).

Falcons changed eyries after successfully raising large broods, a behavior that increased the breeding success of females but not males (Zuberogoitia et al. 2015). This strategy may help mitigate the reduced survival and reproductive chances faced by parents raising large numbers of offspring (Daan et al. 1996). An adaptive decline in average fertility, clutch size, and hatchability with progressively later laying dates is a characteristic feature of bird species that produce a single clutch of variable size per year, such as falcons. Successful individuals exhibit significant variation in productivity, which correlates with the lifespan. Notably, peregrine falcons that start breeding as yearlings tend to achieve greater lifetime reproductive success compared to those entering the breeding pool as adults (Zabala and Zuberogoitia 2015). However, increasing clutch size negatively impacts parental survival (Masman et al. 1989; Dijkstra et al. 1990). Parental daily energy expenditure increases with the number of young in the brood (Xirouchakis et al. 2012). Male parents respond to brood size variation by adjusting their provisioning behavior, whereas female parents adopt a more conservative strategy aimed at maintaining and managing fat reserves collected during the pre-reproductive phase (Wiebe et al. 2000; Xirouchakis et al. 2012). Females also stop hunting during courtship and are fed by males until the young are approximately 2 weeks old (Meijer et al. 1989). Masman et al. (1989) demonstrated that male common kestrels respond strongly to food shortages in the nest by increasing their food delivery rates and flight activity. However, the improvement in hunting efficiency for late-breeding males occurs more slowly than for early breeders (Meijer et al. 1989). Conversely, late-breeding females are more vulnerable to episodes of food shortage due to lower food intake and body mass compared to early breeders (Meijer et al. 1989). Variation in breeding success is largely explained by individual identity, particularly male identity (Zabala and Zuberogoitia 2015). While males adjust their provisioning in response to brood size, females do not exhibit such flexibility (Tolonen and Korpimäki 1994). Instead, female falcons focus on conserving energy reserves, prioritizing their long-term survival over increased provisioning.

Most nestling mortality occurred during the first week after hatching and continued until the chicks were approximately 2 weeks old (Cavé 1968; Wiehn et al. 2000). One contributing factor is food restriction, as evidenced by lower cell-mediated immunity observed in chicks, particularly males, which suggests starvation (Fargallo et al. 2002). Ectoparasites also play a significant role in reducing survival. Nests reused in consecutive years are more likely to become heavily infested with ectoparasites, which can negatively affect both nestlings and adults (McFadzen and Marzluff 1996). Additionally, Piross et al. (2015) observed interannual fluctuations in ectoparasite populations, such as lice, within a breeding population of red-footed falcons, indicating a potential host–parasite interaction.

The post-fledging dependence period is a critical stage for juvenile falcons. During this time, juveniles remain dependent on their parents and stay near the nest site with their family. This phase begins on the day of fledging and continues until the juveniles first leave the natal area. Reproductive performance improves with age in both sexes, but the mechanisms differ. For females, strong selective pressure on noncompetitive breeders drives this improvement. In contrast, males show increased reproductive performance early in life as their hunting skills develop.

7.5.3 Competition

The number of falcon pairs attempting to breed varied from year to year (Cavé 1968; Wiklund 2001). Fluctuations in breeding population size often led to competition among pairs, especially when suitable nesting sites are limited (Cavé 1968; Wiklund and Village 1992; Rodríguez et al. 2007). Nesting site distances among pairs are influenced by the availability and exploitation of favorable food patches (Cavé 1968). When food sources are stable, nesting sites tend to exhibit relatively regular spacing (Solonen 1993). Irregular spacing of nesting places often results from the uneven distribution of preferred habitats and prey (Nilsson et al. 1982; Becker and Sieg 1987; Solonen 1993). Traditionally occupied nest sites have significantly lower variation in productivity, making them higher-quality nesting locations (Wightman and Fuller 2006). In heterogeneous landscapes, these territories are stable in size over time and are defended by pairs year-round, even in cases of mate replacement (Pettifor 1983). The intensity of nest defence by females is closely linked to the age of their young and the brood size (Tolonen and Korpimäki 1994). In contrast, nests defended less aggressively by males show higher rates of desertion and reduced reproductive success (Wiklund 1990, 1996). Predator recognition by prey species is a fundamental trait that enhances survival. Interestingly, the proximity of breeding falcons can provide some protection to prey species from other predators (Wiklund 1979; Quinn et al. 2003; Hipfner et al. 2011). However, the presence of other raptors may serve as a limiting factor for breeding falcons, increasing competition for nesting sites and food while also raising the risk of lethal interactions (Cavé 1968). Shared habitats and prey resources with other raptor species may lead to increased mesopredator suppression, benefiting prey populations (Buchanan 2012).

7.5.4 Feeding Habits and Diet

Falcons possess fine anatomical and physiological adaptations that enable them to capture a variety of prey types. The mean mass of prey correlates positively with the mean body mass of the falcon, highlighting their ability to target appropriately sized prey. Dietary breadth at the species level is influenced by the proportion of birds in the diet, while a higher proportion of invertebrates is negatively correlated with the proportion of mammals consumed (Peery 2000). Mammal-eating falcons demonstrate flexibility by switching to alternative prey when necessary. In contrast, bird-eating falcons must expand their foraging areas to locate adequate avian prey (Korpimäki and Marti 1995; Peery 2000). Prey size and type play a significant role in reinforcing reversed size dimorphism (RSD) in falcons. This pattern becomes more pronounced as falcons transition from feeding on insects to mammals and birds (Slagsvold and Sonerud 2007). Falcons exhibit notable dietary plasticity in response to environmental factors that affect prey distribution and abundance. Seasonal changes significantly impact food composition, prompting falcons to adjust their hunting efforts and techniques accordingly (Shrubb 1982). Habitat heterogeneity is another key factor influencing dietary diversity. Greater habitat complexity increases the number of prey species available, particularly birds, which broadens the food niche of falcons (Peery 2000; Aschwanden et al. 2005; Żmihorski and Rejt 2007).

Falcons adopt hunting techniques based on prey size, prey type, and availability, which are largely shaped by variable environmental factors (Nielsen and Cade 1990; Korpimäki and Marti 1995; Peery 2000). Energy conservation takes precedence over time efficiency during foraging bouts, as falcons aim to minimize energy expenditure (Masman et al. 1988). Avian predators, including falcons, significantly influence the composition of avian and mammalian prey communities within their territories (Suhonen et al. 1994; Ims and Andreassen 2000). Spatial synchrony in rodent population fluctuations is particularly well documented in boreal ecosystems and Central Europe, though these patterns are primarily observed at large spatial scales (Norrdahl and Korpimäki 1993). Falcons that specialize in feeding on migrating small birds face unique challenges. To rear their young, they must migrate to their breeding areas annually and vacate these regions after the breeding season due to the absence of alternative prey outside migration periods (Jany 1960; Vaughan 1961). Prey species employ a range of antipredator strategies, including flocking, vigilance, and mobbing, to counter falcon attacks. Despite these defenses, falcons achieve their highest hunting success when targeting small flocks (Bijlsma and van den Brink 2005; van den Hout et al. 2008; Carere et al. 2009). Populations of specialist predators, such as falcons, often fluctuate in response to the population dynamics of their preferred prey species (Korpimäki and Norrdahl 1991). For example, grouse and gyrfalcon populations exhibit stable, regular, and synchronous 10–12-year quasi-cycles (Nielsen and Pétursson 1995; Brynjarsdottir et al. 2003).

7.5.5 Dispersal Patterns

Dispersal plays a significant role in shaping the distribution of falcon species, as well as the genetic diversity and structure of their populations (Ortego et al. 2008). Across successive breeding seasons, dispersal distances vary both between and within individuals. Female falcons tend to disperse more frequently and exhibit weaker associations with breeding areas compared to males (Wiklund 1996). The proportion of migratory falcons in a population can fluctuate annually, influenced by individual conditions such as age and sex (Lundberg 1988; Kokko 1999). In partially migratory falcon populations, these factors determine whether an individual migrates or remains resident (Thomson 1958; Snow 1968; Serrano and Tella 2003; Vasko et al. 2011; Terraube et al. 2015; Holte et al. 2016). For juvenile falcons, dispersal decisions are critical, as they impact immediate survival and may incur long-term fitness costs (Serrano and Tella 2012).

Both small and large falcon species display high rates of philopatry, with individuals often returning to their natal areas. However, juveniles are more likely to disperse over longer distances compared to adults (Mearns and Newton 1984; Serrano and Tella 2003; Serrano et al. 2008; Burnham and Newton 2011). Before migrating, social falcons gather in large communal roosts, where they exploit available food resources, an important behavior for preparing for migration (García Fernández 2000; Barriocanal et al. 2004; De Frutos et al. 2007; Fehérvári et al. 2014; Palatitz et al. 2015; Strucker and Luitwieler 2015).

Natural barriers, such as large water bodies, dense forests, and deserts, pose significant challenges for migrating falcons (Meyer et al. 2000; Strandberg et al. 2009a, b). Despite these obstacles, falcons are capable of covering long distances daily and exhibit frequent nocturnal migration to optimize their journeys (Limiñana et al. 2012). Falcons aim to balance energy expenditure, the distance to nonbreeding areas, and constraints encountered along their routes. These constraints include both predictable obstacles, such as natural barriers, and unpredictable factors, like adverse weather conditions (Mellone et al. 2013).

7.5.6 Gregariousness and Coloniality

A few falcon species, such as lesser kestrels, red-footed falcons, and Eleonora's falcons, exhibit gregarious behavior (Clark 1999). Among these, lesser kestrels and Eleonora's falcons are colonial and migratory birds, with breeding pairs either settling in pre-existing colonies or establishing new ones. Living in larger colonies provides significant fitness benefits, primarily through increased survival probabilities for adults and nestlings due to reduced predation rates (Serrano et al. 2001, 2005). Despite these benefits, colonial living imposes notable costs, including density-dependent food depletion in larger colonies (Bonal and Aparicio 2008). Nevertheless, most lesser kestrels and Eleonora's falcons return to the same breeding colony where they attempted to breed the previous year (Ristow et al. 1991; Negro et al. 1997; Calabuig et al. 2008b). Colony growth is strongly linked to breeding success in the previous year, which supports the continued prevalence of colonial behavior in these species (Serrano et al. 2004, 2005).

The presence of conspecifics is a key indicator of colony quality. Falcons are more likely to move to larger colonies rather than smaller ones (Vaughan 1961;

Serrano et al. 2001, 2005; Serrano and Tella 2003). However, dispersing individuals prioritize highly productive colonies—those with greater numbers of successful pairs and fledglings—over merely large colonies (Aparicio et al. 2007; Calabuig et al. 2008a, b).

Mixed-species assemblages during breeding seasons provide additional advantages. These assemblages enhance reproductive success by reducing predation pressure from common predators (Blanco and Tella 1997; Campobello et al. 2012).

7.5.7 Conservation

The conservation, effective management, and mitigation of threats to falcon populations in the Western Palearctic have become increasingly urgent. Over the past two decades, field studies have focused on identifying and addressing a broader range of threats, which now extend beyond historical pressures such as persecution and human disturbance. Key challenges include habitat alteration in both natural and agricultural landscapes, the proliferation of wind and solar farms, and the rapid expansion of mass tourism. Tackling these interconnected issues is essential to ensuring the long-term survival of these species.

7.5.7.1 Population Recovery of Falcons Since the Pesticide Ban

The recovery of breeding populations, particularly peregrine falcons, has been significantly influenced by the banning of chemical compounds such as DDT (Kéry et al. 2022). This recovery, however, was gradual, spanning four decades, with varying rates of improvement across different measures of reproductive performance, including egg, nestling, and fledgling production (Oli et al. 2023). Despite these gains, survival rates remained generally lower across all age classes due to the chronic and persistent effects of organochlorine pesticides during the recovery process (Smith et al. 2015). Notably, the resurgence became most evident in the 1990s, when the frequency of deaths attributed to organochlorine poisoning began to decline significantly (Newton et al. 1999; Horne and Fielding 2002). Research indicates that reproductive success in territories previously occupied during the breeding season has steadily increased over time (McGrady et al. 2017; Oli et al. 2023). This evidence aligns with the hypothesis that the phased reduction and eventual ban of organochlorine pesticides enhanced the reproductive success of falcons. However, the recovery has shown regional variations, suggesting that the impact of these pesticides differed geographically and may have been compounded by other factors, such as persecution (McGrady et al. 2017). The ban also had wider-reaching effects. For example, the newly established population was heavily influenced by the growth of the German population and the overall expansion of the European population as a whole (Beran et al. 2009). Despite these differences, the peregrine falcon is no longer considered a threatened species. It has been reclassified as Near Threatened on the Red List (Gjershaug et al. 2008; Jacobsen et al. 2008). In Great Britain, populations have rebounded impressively, growing from a low of 118 pairs in the 1960s to nearly 2000 pairs today—a record high (Ratcliffe 1963; Wilson et al. 2018).

The impact of chemical compounds has, in some cases, been negligible, depending on the territorial context. For instance, no organochlorine (OC) or polychlorinated biphenyl (PCB) residues were found in lanner falcon chick blood above the detection limits (Movalli et al. 2008). Similarly, low pollutant concentrations in lesser kestrels suggest that these birds either ranged across unpolluted areas throughout the year or were less affected due to their short food chain (Goutner et al. 2015).

Ultimately, no genetic imprint of the population bottleneck experienced by peregrine falcons has been detected (Ponnikas et al. 2017). This is likely because there has not been enough time for genetic erosion to occur due to the bottleneck (Weaving et al. 2021). Interestingly, in the UK, there were no formal reintroduction programs for peregrine falcons, and it seems the population has recovered from its crash without human intervention (Weaving et al. 2021). Similar to the Czech Republic population, high dispersal rates are thought to have contributed to maintaining genetic diversity across the landscape and preventing negative genetic effects (Ponnikas et al. 2017).

7.5.7.2 ISSAPs

The International Single Species Action Plan (ISSAP) is designed to protect threatened or endangered species through coordinated conservation efforts across multiple countries or regions. It outlines specific actions needed to address threats, improve habitats, and support species recovery. The plan includes an assessment of the species' status and prioritizes collaboration among stakeholders for effective implementation. Monitoring and evaluation ensure the progress of conservation actions and the species' recovery.

Several ISSAPs have been developed and implemented for species within the *Falco* genus, including those for Eleonora's falcon (Birdlife International 1999), saker falcon (Kovács et al. 2014), lanner falcon (Gustin et al. 2002), gyrfalcon (Koskimies 2005), lesser kestrel (Biber 1996; Iñigo and Barov 2010), and the recent one for the sooty falcon (Leonardi et al. 2024a). In certain cases, regional plans have also been created, particularly for species with important locally nesting populations. Notable examples include the Italian plans for the lanner falcon (Andreotti and Leonardi 2007) and Eleonora's falcon (Spina and Leonardi 2007).

7.5.7.3 Conservation Measures for Falcon Populations

The conservation and protection of Western Palearctic falcons are governed by several international and regional agreements and conventions (Table 7.1). The inclusion of all falcons under CITES (Convention on International Trade in Endangered

Table 7.1 International and regional agreements and conventions that include falcons

Species	CITES	CMS	Raptors MOU	EU birds directive
F. naumanni	Appendix II	Appendix I and II	Category 2	Annex I
F. tinnunculus	Appendix II	Appendix II	Category 2	–
F. vespertinus	Appendix II	Appendix I and II	Category 1	Annex I
F. eleonorae	Appendix II	Appendix II	Category 3	Annex I
F. concolor	Appendix II	Appendix II	Category 1	–
F. columbarius	Appendix II	Appendix II	Category 3	Annex I
F. subbuteo	Appendix II	Appendix II	Category 2	–
F. biarmicus	Appendix II	Appendix II	Category 2	Annex I
F. cherrug	Appendix II	Appendix I and II	Category 1	Annex I
F. rusticolus	Appendix I	Appendix II	Category 3	Annex I
F. peregrinus	Appendix I	Appendix II	Category 3	Annex I

CITES includes species in Appendix I, which are threatened with extinction and may be affected by trade, and Appendix II, which covers species that are not necessarily threatened with extinction but could become so if their trade is not regulated. The CMS (Convention on the Conservation of Migratory Species of Wild Animals) includes species in Appendix I, which are endangered, and Appendix II, which lists species with an unfavorable conservation status or that would benefit significantly from international cooperation through agreements. The Raptors MoU, a Memorandum of Understanding under CMS focused on the conservation of migratory birds of prey in Africa and Eurasia, identifies Category 1 species as those globally threatened with extinction, Category 2 species as those not globally threatened but requiring concerted conservation efforts due to potential risks or declining trends, and Category 3 species as those neither globally threatened nor with an unfavorable conservation status. Finally, the EU Birds Directive includes Annex I, which lists bird species that are particularly rare or vulnerable and therefore require special conservation measures

Species of Wild Fauna and Flora) is particularly important due to the continued trade of these species for falconry and as pets (Table 7.1; see also Chaps. 4, 5, and 6).

In recent decades, numerous conservation efforts have been directed toward individual species, with a primary emphasis on those most endangered (Tables 7.1 and 7.2). Conservation efforts for the lesser kestrel, initiated in the 1980s, have prioritized a few key actions. These include preserving colony sites by regulating building renovations, creating new nesting opportunities through artificial nest installations, and implementing breeding and release programs in selected areas (Bux et al. 2008; Orta and Kirwan 2020; Yaneva et al. 2022). In Spain, an effective mitigation measure has been implemented by tilling the soil around the base of wind turbines, especially in areas with high collision rates (Pescador et al. 2019). From 2013 to 2015, a release program successfully re-established the species as a breeder in Bulgaria (Gradev et al. 2016). At least five European countries have implemented systematic breeding programs for this species (Iñigo and Barov 2010; Derlink et al. 2018). While the common kestrel is not considered an endangered species, breeding schemes have been established in several European countries (Derlink et al. 2018). Additionally, the widespread use of nest boxes by this small falcon offers valuable data on behavioral ecology, physiology, and parasitology, contributing to the conservation of all species in the genus *Falco* (Fargallo et al. 2001; Costantini et al.

2007; Kasprzykowski et al. 2021). Given its vulnerable conservation status, mid-term conservation measures for the red-footed falcon have included the creation of artificial colonies using nest boxes (Palatitz et al. 2015). This action seems to be effective in Italy and Slovakia in terms of occupancy and productivity, as it has slowed the decline in numbers (Slobodnik et al. 2017; Calabrese et al. 2020). As a result, it could be replicated in other territorial contexts (Fehérvári et al. 2012). Efforts to conserve Eleonora's falcon are mainly centered around monitoring key nesting colonies in Greece, Spain, Cyprus, Tunisia, and Morocco (Baron and Derhé 2011; Qninba et al. 2016). In Morocco, the species has notably expanded, a result attributed to the cessation of direct persecution (Rguibi et al. 2012). According to the recent ISSAP for the sooty falcon, our understanding of the breeding population is still limited, particularly in inland desert areas (Leonardi et al. 2024a). One of the key priorities moving forward is to carry out local surveys to provide a more accurate population estimate. Research conducted on island colonies has offered important insights into the survival and mortality rates of both juveniles and adults, especially during migration (McGrady et al. 2016). Furthermore, expanding conservation actions to wintering grounds in Madagascar and migration routes in Eastern Africa is crucial, especially in the context of Locust and Quelea control operations (Leonardi et al. 2024a). Monitoring of the merlin continues to assess its potential exposure to environmental pollutants, such as organochlorines and mercury (Walker et al. 2011). Efforts focus on preserving a mix of deciduous or coniferous tree groves for nesting and open grasslands for hunting, according to the species' habitat preferences and breeding site requirements (Konrad et al. 2020). Like the merlin, the Eurasian hobby is the focus of numerous breeding programs across several European countries (Derlink et al. 2018).

Among the large falcons, the lanner falcon is one of the most concerning species in terms of status and distribution (see Chaps. 4, 5, and 6; Tables 7.1 and 7.2; Leonardi 2015). A recent comprehensive paper has shown how the species was resilient, with its fragmented population in the past restricted to glacial refugia in Southern Europe and North Africa (Leonardi et al. 2024b). The subspecies

Table 7.2 Estimated population size of Western Palearctic falcons, along with their classification in the Red List categories

Species	Red List category	Population size (bps)
F. naumanni	Least concern	32,900–42,600
F. tinnunculus	Least concern	411,000–631,000
F. vespertinus	Vulnerable	57,800–84,800
F. eleonorae	Least concern	14,200–14,500
F. concolor	Vulnerable	1400–2000
F. columbarius	Least concern	20,000–41,700
F. subbuteo	Least concern	140,000–219,000
F. biarmicus	Least concern	420–850
F. cherrug	Endangered	430–630
F. rusticolus	Least concern	1000–1800
F. peregrinus	Least concern	16,100–31,100

Source: https://datazone.birdlife.org/

Fig. 7.5 Different steps of the LIFE project in Italy, focusing on the captive breeding, training, and release of lanner falcons into the wild. (Credit: V. Cippitelli)

feldeggii is facing a significant decline, while the other two subspecies appear to be out of immediate danger (IUCN and DEF 2020; De Santis et al. 2023). However, in some inland areas of the Middle East, populations are considered close to extinction (Alabdulhafith et al. 2022). The species has also been a target of various LIFE projects, the latest focusing on breeding and training in captivity, followed by release into the wild (Fig. 7.5).

The saker falcon is classified as vulnerable, with the nesting populations in Europe being particularly at risk (Tables 7.1 and 7.2). In response, new research initiatives are gathering crucial baseline data on their distribution, population, ecology, and threats. One method being employed is satellite tracking, which helps to map individual movements and site usage (Bagyura et al. 2012; Gamauf and Dosedel 2012; Prommer et al. 2012). These conservation efforts have already shown positive results, with population trends improving across Europe (Dixon 2012). In Bulgaria, a captive-breeding and release program has been introduced to restore the extirpated breeding population of saker falcons. Furthermore, installing breeding platforms on electric pylons has been shown to significantly boost breeding success (Beran et al. 2009). Between 2015 and 2020, several falcons were released, and by 2018, the first successful breeding in the wild was confirmed (Lazarova et al. 2021). In Hungary, conservation initiatives have been highly successful, increasing the breeding population from approximately 30 pairs in 1980 to about 240 pairs by 2010 (Bagyura et al. 2012). Moreover, a pair of saker falcons that bred in Bulgaria in 2018 and 2019 were from a cohort released in 2015, demonstrating that the translocation of captive-bred birds could be an effective strategy for restoring the species to its historical breeding range (Dixon et al. 2019). Gyrfalcon populations in Europe seem to be more stable. Breeding programs are being carried out in nearly all countries within the species' range (Derlink et al. 2018). However, given their Arctic boreal distribution, further research is needed to understand the potential effects of global warming on the species (Matz et al. 2011; Franke 2017). Efforts to fully restore peregrine falcon populations across their former range, which includes Germany, Poland, Russia, Belarus, and the Baltic States, still require significant attention (Sielicki and Sielicki 2009; Kleinstäuber et al. 2018). In southern Scandinavia, a reintroduction project using non-native genetic stock played a crucial role in preventing the species from going extinct. This initiative helped preserve genetic diversity and reduce the risk of inbreeding depression (Jacobsen et al. 2008). Currently, the peregrine falcon is classified as Least Concern (LC), as its population does not meet the criteria for Vulnerable status (Tables 7.1 and 7.2).

A key challenge for the future will be ensuring population estimates are both highly reliable and easily comparable (Table 7.2). These estimates will play a critical role in shaping conservation efforts, guiding actions at both global and local levels. Moreover, they will help assess the potential population growth expected once these initiatives are implemented. Despite the many breeding schemes conducted each year, there is still a significant knowledge gap regarding rare and geographically fragmented species like the lanner falcon and the sooty falcon (Leonardi 2015; Leonardi et al. 2024a). Their population estimates often remain inconsistent and difficult to compare.

7.6 Conclusions

Over the past 80 years, following the Second World War, falcon populations in the Western Palearctic have faced significant declines. This deterioration is largely attributed to the systematic persecution of the nineteenth and early twentieth centuries, followed by the subsequent widespread use of pesticides. The almost complete recovery of several species has been achieved through comprehensive legal protection at both international and local levels, alongside the implementation of various conservation measures such as reintroduction, captive breeding, nest boxes, habitat maintenance, and field research. The biology, ecology, and ethology of falcons have never been studied as extensively in contemporary history. As a result, an increasing number of papers are being published in a wide range of ecology journals. New threats have emerged with the expansion of human activities, including power lines, wind and solar farms, and changes in agricultural practices.

Over the approximately 1,800,000 years from the Early Pleistocene to the Neolithic period, the distribution and consistency of Western Palearctic falcons have undergone minimal changes, with variations occurring only on a local scale (see figures in "*Preface*"). This is because all falcon species exhibit strong climatic niche conservatism, which ultimately limits their geographical range. In this context, it is essential to assess the potential impact of increasing global warming, alongside the growing pressure from human activities.

References

Alabdulhafith B, Binothman A, Alwahiby A, Haig SM, Prommer M, Leonardi G (2022) Predicting the potential distribution of a near-extinct avian predator on the Arabian Peninsula: implications forits conservation management. Environ Monit Assess 194:535. https://doi.org/10.1007/s10661-022-10225-2

Amato M, Ossino A, Brogna A et al (2014) Influence of habitat and nest-site quality on the breeding performance of Lanner Falcons Falco biarmicus. Acta Ornithol 49:1–7. https://doi.org/10.3161/000164514X682841

Andreotti A, Leonardi G (2007) Italian Action Plan for the Lanner Falcon (*Falco biarmicus feldeggii*). Quad Cons Natura No. 24, Min Ambiente and INFS, Rome

Aparicio JM, Bonal R, Muñoz A (2007) Experimental test on public information use in the colonial lesser kestrel. Evol Ecol 21:783–800. https://doi.org/10.1007/s10682-006-9151-7

Aschwanden J, Birrer S, Jenni L (2005) Are ecological compensation areas attractive hunting places for kestrels *Falco tinnunculus* and long-eared owls *Asio otus*? J Ornithol 146:279–286. https://doi.org/10.1007/s10336-005-0090-9

Attili L, Garofalo L, Puddu G, Tirone G, Pizzarelli A, Barbara N, Haring E, Lorenzini R (2023) Genetic distinctiveness of an endangered falcon: implications for conservation in Europe. PLoS ONE 18:e0295424. https://doi.org/10.1371/journal.pone.0295424

Bagyura J, Szitta T, Haraszthy L, Viszló L, Fidlóczky J, Prommer M (2012) Results of the Saker Falcon (*Falco cherrug*) conservation programme in Hungary between 1980–2010. Aquila 119:105–110

Baker JR (1938) The evolution of breeding seasons. In: de Beer GR (ed) Evolution, essays on aspects of evolutionary biology. Clarendon Press, Oxford, pp 161–177

Baron B, Derha M (2011) Review of the implementation of species action plans of threatened birds in the European Union (2004–2010). BirdLife International for the European Commission, Bruxelles

Barriocanal C, Burgas D, Salvà C, Camps J (2004) Unusual concentration of the red-footed falcon *Falco vespertinus* in the spring migration in NE Spain. Scientia Gerundensis 27:31–34

Becker DM, Sieg CH (1987) Home range and habitat utilisation of breeding male Merlins, *Falco columbarius*, in southeastern Montana. Can Field Nat 101:398–403

Beran V, Hlaváč V, Bělka T (2009) Development of the peregrine and saker falcon populations in the Czech Republic in the last 30 years. In: Sielicky J, Mizera T (eds) Peregrine falcon populations—status and perspectives in the 21st century. Turul—Poznań University of Life Sciences Press, Warsaw–Poznań, pp 757–764

Bergman G (1961) The food of birds of prey and owls in Fenno-Scandia. Br Birds 54:307–320
Biber J-P (1996) International Action Plan for the Lesser Kestrel (*Falco naumanni*). BirdLife International, Cambridge
Bijleveld M (1974) Birds of prey in Europe. Macmillan Press, London
Bijlsma R, van den Brink B (2005) A barn swallow *Hirundo rustica* roost under attack: timing and risks in the presence of African Hobbies *Falco cuvieri*. Ardea 93:37–48
Birdlife International (1999) International Species Action Plan: Eleonoras' s falcon *Falco eleonorae*. D. Ristow, compiler, unpublished data. http://europa.eu.int/comm/environment/nature/directive/birdspriority.htm
Blanco G, Tella JL (1997) Protective association and breeding advantages of choughs nesting in lesser kestrel colonies. Anim Behav 54:335–342. https://doi.org/10.1006/anbe.1996.0465
Blondel J (1967) Réflexions sur les rapports entre prédateurs et proies chez les Rapaces. I. Les effets de la prédation sur les populations de proies. La Terre et La Vie 21:5–32
Bogan JA, Mitchell J (1973) Continuing dangers of peregrines from dieldrin. Br Birds 66:437–439
Bonal R, Aparicio JM (2008) Evidence of prey depletion around lesser kestrel *Falco naumanni* colonies and its short term negative consequences. J Avian Biol 39:189–197. https://doi.org/10.1111/j.2008.0908-8857.04125.x
Bonora M, Chiavetta M (1975) Contribution à l'étude du Faucon lanier Falco biarmicus feldeggii en Italie. Nos Oiseaux 33:153–168
Booth BD (1961) Breeding of the Sooty Falcon in the Libyan desert. Ibis 103:129–130
Bruggeman JE, Swem T, Andersen DE, Kennedy PL, Nigro D (2016) Multi-season occupancy models identify biotic and abiotic factors influencing a recovering Arctic Peregrine falcon *Falco peregrinus tundrius* population. Ibis 158:61–74. https://doi.org/10.1111/ibi.12313
Brynjarsdottir J, Lund SH, Magnusson KG, Nielsen OK (2003) Analysis of time series for rock ptarmigan and gyrfalcon populations in north-east Iceland. http://raunvisindastofnun.hi.is/sites/raunvisindastofnun.hi.is/files/rh-18-2003.pdf
Buchanan JB (2012) Change in merlin hunting behaviour following recovery of peregrine falcon population suggests mesopredator suppression. J Raptor Res 46:349–456. https://doi.org/10.3356/JRR-11-75.1
Burke BJ, Clarke D, Fitzpatrick A, Carnus T, McMahon BJ (2015) Population status and factors affecting the productivity of peregrine falcon *Falco peregrinus* in County Wicklow, Ireland, 2008–2012. Biol Environ 115:115–124. https://doi.org/10.1353/bae.2015.0016
Burnham KK, Newton I (2011) Seasonal movements of Gyrfalcons *Falco rusticolus* include extensive periods at sea. Ibis 153:468–484. https://doi.org/10.1111/j.1474-919X.2011.01141.x
Bux M, Giglio G, Gustin M (2008) Nest box provision for lesser kestrel *Falco naumanni* populations in the Apulia region of southern Italy. Cons Evidence 6:58–61
Cade TJ (1960) Ecology of the peregrine and gyrfalcon populations in Alaska. Univ Calif Publ Zool 63:151–290
Cade TJ, Weaver JD, Platt JB, Burnham WA (1977) The propagation of large falcons in captivity. Rap Res 11:28–48
Calabrese L, Mucciolo A, Zanichelli A, Gustin M (2020) Effects of nest boxes on the most important population of red-footed falcon Falco vespertinus in Italy. Cons Evidence 20:55–39
Calabuig G, Ortego J, Aparicio JM, Cordero PJ (2008a) Public information in selection of nesting colony by lesser kestrels: which cues are used and when are they obtained? Anim Behav 75:1611–1617. https://doi.org/10.1016/j.anbehav.2007.10.022
Calabuig G, Ortego J, Cordero PJ, Aparicio JM (2008b) Causes, consequences and mechanisms of breeding dispersal in the colonial lesser kestrel, Falco naumanni. Anim Behav 76:1989–1996. https://doi.org/10.1016/j.anbehav.2008.08.019
Campobello D, Sará M, Hare JF (2012) Under my wing: lesser kestrels and jackdaws derive reciprocal benefits in mixed-species colonies. Behav Ecol 23:425–433. https://doi.org/10.1093/beheco/arr207

Carere C, Montanino S, Moreschini F, Zoratto F, Chiarotti F, Santucci D et al (2009) Aerial flocking patterns of wintering starlings, *Sturnus vulgaris*, under different predation risk. Anim Behav 77:101–107. https://doi.org/10.1016/j.anbehav.2008.08.034

Cavé AJ (1968) The breeding of the kestrel, Falco tinnunculus L., in the reclaimed area Oostelijk Flevoland. Netherland J Zool 18(3):313–340. https://doi.org/10.1163/002829668X00027

Chavko J, Slobodník R, Deutschová L, Lipták J, Mihók J, Obuch J, Nemček V (2014) The Saker Falcon (*Falco cherrug*) population, diet and nest boxes in Slovakia: LIFE-project report 2011–2014. Slovak Raptor J 8:73–86

Chiavetta M, Bonora M (1973) Le faucon lanier Falco biarmicus niche en Italie du Nord. Nos Oiseaux 32:23

Christensen S, Nielsen BP, Porter RF, Willis I (1973a) Flight identification of European raptors. Part 6 Large falcons. Br Birds 66:100–114

Christensen S, Nielsen BP, Porter RF, Willis I (1973b) Flight identification of European raptors. Part 7 Eleonora's falcon, hobby, red-footed falcon and merlin. Br Birds 66:285–298

Christensen S, Nielsen BP, Porter RF, Willis I (1973c) Flight identification of European raptors. Part 8 kestrels, accipiters, and black-winged kite. Br Birds 66:472–493

Clark WS (1999) A field guide to the raptors of Europe, the Middle East, and North Africa. Oxford University Press, Oxford

Clark AL, Peakall DB (1977) Organochlorine residues in Eleonora's falcon *Falco eleonorae*, its eggs and its prey. Ibis 119:353–358. https://doi.org/10.1111/j.1474-919X.1977.tb08254.x

Conder P (1977) Legal status of birds of prey and owls in Europe. Proc ICBP World Conf Birds Prey Vienna 1975:189–193

Costantini D, Casagrande S, Dell'Omo G (2007) MF magnitude does not affect body condition, pro-oxidants and anti-oxidants in Eurasian kestrel (*Falco tinnunculus*) nestlings. Environ Res 104:361–366. https://doi.org/10.1016/j.envres.2007.02.006

Cramp S (1963) Toxic chemicals and birds of prey. Brit Birds 53:124–139

Cramp S, Simmons KEL (1980) The birds of the western palearctic, vol 2. Oxford University Press, Oxford

Csermely D (1993) Duration of the rehabilitation period and familiarity with the prey affect the predatory behavior of captive wild kestrel (*Falco tinnunculus*). Ital J Zool 60:211–214. https://doi.org/10.1080/11250009309355812

Curry-Lindahl K (1961a) Conservation and predation problems of birds of prey in Sweden. Br Birds 54:297–306

Curry-Lindahl K (1961b) Landscape changes and the vertebrate fauna in Sweden during the last 150 years. Bijdr td Dierkunde 31:27–44

Curry-Lindahl K (1970) Bird conservation in Europe. Bird Study 17:61–80. https://doi.org/10.1080/00063657009476260

Daan S, Deerenberg C, Dijkstra C (1996) Increased daily work precipitates natural death in the kestrel. J Anim Ecol 65:539–544. https://doi.org/10.2307/5734

De Frutos Á, Olea PP, Vera R (2007) Analyzing and modelling spatial distribution of summering lesser kestrel: the role of spatial autocorrelation. Ecol Model 200:33–44. https://doi.org/10.1016/j.ecolmodel.2006.07.007

De Lucca C (1969) Bird migration over the Maltese Islands. Ibis 111:322–337. https://doi.org/10.1111/j.1474-919X.1969.tb02548.x

De Santis E, Leonardi G, Jennings V, Catoni C, Sarrocco S, Sabatini G, Puddu G (2023) A new snapshot of the Italian breeding population of Lanner falcon *Falco biarmicus feldeggii*. Poster at the 21th Italian Ornithological Congress, Varese

Demandt C (1953) Brutbiologische probleme beim Wanderfalken (*Falco peregrinus*). J Ornithol 94:99–102. https://doi.org/10.1007/BF01950833

Dementiev GP, Gortchakovskaya NN (1945) On the biology of the Norwegian gyrfalcon. Ibis 87:559–565. https://doi.org/10.1111/j.1474-919X.1945.tb01387.x

Derlink M, Wernham C, Bertoncelj I, Kovács A, Saurola P, Duke G, Movalli P, Vrezec A (2018) A review of raptor and owl monitoring activity across Europe: its implications for capacity

building towards pan-European monitoring. Bird Study 65:S4–S20. https://doi.org/10.1080/00063657.2018.1447546
Dijkstra C, Bult A, Bijlsma S, Daan S, Meijer T, Zijlstra M (1990) Brood size manipulations in the kestrel (*Falco tinnunculus*): effects on offspring and parent survival. J Anim Ecol 59:269–285. https://doi.org/10.2307/5172
Dixon A (2012) Conservation of the Saker Falcon (*Falco cherrug*) and the use of hybrids for falconry. Aquila 119:9–19
Dixon A, Ragyov D, Izquierdo D, Weeks D, Rahman ML, Klisurov I (2019) Movement and survival of captive-bred Saker Falcons *Falco cherrug* released by wild hacking: implications for reintroduction management. Acta Ornithol 54:157–170. https://doi.org/10.3161/00016454AO2019.54.2.003
Donázar JA, Cortés-Avizanda A, Fargallo JA, Margalida A, Moleón M, Morales- Reyes Z et al (2016) Roles of raptors in a changing world: from flagships to providers of key ecosystem services. Ardeola 63:181–234. https://doi.org/10.13157/arla.63.1.2016.rp8
Erdélyi K, Tenk M, Dan A (1999) Mycoplasmosis associated perosis type skeletal deformity in a saker falcon nestling in Hungary. J Wildl Dis 35:586–590. https://doi.org/10.7589/0090-3558-35.3.586
Falk K, Møller S, Rigét FF, Sørensen PB, Vorkamp K (2018) Raptors are still affected by environmental pollutants: Greenlandic peregrines will not have normal eggshell thickness until 2034. Ornis Hung 26:171–176. https://doi.org/10.1515/orhu-2018-0026
Fargallo JA, Blanco G, Potti J, Viñuela J (2001) Nestbox provisioning in a rural population of Eurasian kestrels: breeding performance, nest predation and parasitism. Bird Study 48:236–244. https://doi.org/10.1080/00063650109461223
Fargallo JA, Laaksonen T, Pöyri V, Korpimäki E (2002) Inter-sexual differences in the immune response of Eurasian kestrel nestlings under food shortage. Ecol Lett 5:95–101. https://doi.org/10.1046/j.1461-0248.2002.00290.x
Fehérvári P, Solt S, Barna K, Agoston A, Gergely J, Nagy A, Harnos A (2012) Allocating active conservation measures using species distribution models: a case study of red-footed falcon breeding site management in the Carpathian Basin. Anim Conserv 15:648–657. https://doi.org/10.1111/j.1469-1795.2012.00559.x
Fehérvári P, Lázár B, Palatitz P, Solt S, Nagy A et al (2014) Pre-migration roost site use and timing of post-nuptial migration of Red-footed falcons (Falco vespertinus) revealed by satellite tracking. Ornis Hung 22:36–47. https://doi.org/10.2478/orhu-2014-0009
Franke A (2017) Priorities for gyrfalcon research: food, weather, and phenology in a changing climate. In: Anderson DL, McClure CJW, Franke A (eds) Applied raptor ecology: Essentials from gyrfalcon research. The Peregrine Fund, Boise, pp 11–32. https://doi.org/10.4080/are.2017/001
Frochot J (1967) Réflexions sur les rapports entre prédateurs et proies chez les Rapaces. II. l'influence des proies sur les rapaces. La Terre et La Vie 21:33–62
Frumkin R, Pinshow B (1983) Notes on the breeding ecology and distribution of the Sooty Falcon *Falco concolor* in Israel. Ibis 125:251–259. https://doi.org/10.1111/j.1474-919X.1983.tb03108.x
Gahbauer MA, Bird DM, Armstrong TR (2015) Origin, growth and composition of the recovering Peregrine Falcon population in Ontario. J Raptor Res 49:281–293. https://doi.org/10.3356/JRR-14-00002.1
Gamauf A, Dosedel R (2012) Satellite telemetry of Saker Falcons (*Falco cherrug*) in Austria: juvenile dispersal at the westernmost distribution limit of the species. Aquila 119:65–78
García Fernández JC (2000) Dispersión premigratoria del Cernícalo Primilla *Falco naumanni* en España. Ardeola 47:197–202
García-Fernández AJ, Motas-Guzmán M, Navas I et al (1997) Environmental exposure and distribution of lead in four species of raptors in Southeastern Spain. Arch Environ Contam Toxicol 33:76–82. https://doi.org/10.1007/s002449900226
Géroudet P (1969) Les espèces d'Oiseaux menacées en Europe. Biol Conserv 1:213–217. https://doi.org/10.1016/0006-3207(69)90148-7

Gilbertson M (1969) The distribution of the peregrine falcon in Northern Ireland. Irish Nat J 16:131–133

Gjershaug JO, Kålås JA, Nygård T, Herzke D, Folkestad AO (2008) Monitoring of raptors and their contamination levels in Norway. AMBIO J Hum Environ 37:420–424. https://doi.org/10.1579/0044-7447(2008)37[423:MORATC]2.0.CO;2

Glading B, Tillotson DF, Selleck DM (1943) Raptor pellets as indicators of food habits. Calif Fish Game 29:92–121

Glutz Von Blotzheim UN, Bauer KM, Bezzel E (1971) Handbuch der Vögel Mittleuropas, vol 4. Akademische Verlegsgesellschaft, Frankfurt am Main

Goutner V, Bakaloudis DE, Papakosta MA, Vlachos CG, Mattig FR, Pijanowska U, Becker PH (2015) Organochlorine and mercury residues in eggs of the lesser kestrel (Falco naumanni) from a long term study in the eastern Mediterranean. Environ Pollut 207:196–204. https://doi.org/10.1016/j.envpol.2015.09.021

Gradev G, Marin S, Zhelev P, Antolin J (2016) Recovering the Lesser kestrel (*Falco naumanni*) as a breeder in Bulgaria. In: First national conference of reintroduction of conservation-reliant species, University Press, pp 136–144

Greenwood A (1977) The role of disease in the ecology of British raptors. Bird Study 24:259–265. https://doi.org/10.1080/00063657709476565

Griffiths ME (1967) The population density of the Kestrel in Leicestershire. Bird Study 14:184–190. https://doi.org/10.1080/00063656709476161

Guichard KM (1957) The spring migration in Tripolitania—1955. Ibis 99:106–114. https://doi.org/10.1111/j.1474-919X.1957.tb01936.x

Gustin M, Palumbo G, Corso A (2002) International Species Action Plan for the Lanner Falcon *Falco biarmicus*. BirdLife International, Council of Europe, T-PVS/Inf, 16, Strasbourg, 20 pp

Hackett SJ, Kimball RT, Reddy S et al (2008) A phylogenomic study of birds reveals their evolutionary history. Science 320(5884):1763–1768. https://doi.org/10.1126/science.1157704

Hipfner JM, Morrison KW, Darvill R (2011) Peregrine falcons enable two species of colonial seabirds to breed successfully by excluding other aerial predators. Waterbirds 34:82–88. https://doi.org/10.1675/063.034.0110

Hiraldo F, Negro JJ, Donazar JA, Gaona P (1996) A demographic model for a population of the endangered lesser kestrel in southern Spain. J Appl Ecol 33:1085–1093. https://doi.org/10.2307/2404688

Holt G, Frøslie A, Norheim G (1979) Mercury, DDE, and PCB in the avian fauna in Norway 1965–1976. Acta Vet Scand Suppl 70:1–28

Holte D, Köppen U, Schmitz-Ornés A (2016) Partial migration in a Central European raptor species: an analysis of ring re-encounter data of common kestrels *Falco tinnunculus*. Acta Ornithol 51:39–54. https://doi.org/10.3161/00016454AO2016.51.1.004

Horne G, Fielding AH (2002) Recovery of the peregrine falcon Falco peregrinus in Cumbria, UK, 1966–99. Bird Study 49:229–236. https://doi.org/10.1080/00063650209461270

Horváth L (1955) The life of the red-legged falcon (*Falco vespertinus*) in the Ohat forest. Acta Congr Int Orn 11:583–587

Horváth L (1975) Social pattern and behaviour between two *Falco* species (*Aves*). Ann Hist-Nat Mus Nat Hung 67:327–331

Hunt WG (1998) Raptor floaters at. Moffat's equilibrium. Oikos 82:191–197. https://doi.org/10.2307/3546929

Ims RA, Andreassen HP (2000) Spatial synchronization of vole population dynamics by predatory birds. Nature 408:194–196. https://doi.org/10.1038/35041562

Iñigo A, Barov B (2010) Action plan for the lesser kestrel *Falco naumanni* in the European Union. SEO|BirdLife and BirdLife International for the European Commission

IUCN and DEF (2020) Premier dénombrement national des rapaces rupicoles diurnes du Maroc. Résultats 2019. Malaga, Espagne et Rabat, Maroc UICN et DEF

Jacobsen F, Nesje M, Bachmann L et al (2008) Significant genetic admixture after reintroduction of peregrine falcon (*Falco peregrinus*) in Southern Scandinavia. Conserv Genet 9:581–591. https://doi.org/10.1007/s10592-007-9373-4

Jany E (1960) An Brutplätzen des Lannerfalken (*Falco biarmicus erlangeri* Kleinschmidt) in einer Kieswüste der inneren Sahara (Nordrand des Serir Tibesti) zur Zeit des Frühjahrszugs. Proc Int Ornithol Congr 21:343–352

Jarvis ED, Mirarab S, Aberer AJ et al (2014) Whole-genome analyses resolve early branches in the tree of life of modern birds. Science 346(6215):1320–1331. https://doi.org/10.1126/science.1253451

Kasprzykowski Z, Rzępała M, Golawski A (2021) The effect of local weather conditions and nest box location on the reproduction of the Common Kestrel (*Falco tinnunculus*) in the farmland of eastern Poland. Ornis Fennica 98:97–104. https://doi.org/10.51812/of.133987

Kéry M, Banderet G, Müller C, Pinaud D, Savioz J, Schmid H, Werner S, Monneret RJ (2022) Spatio-temporal variation in post-recovery dynamics in a large Peregrine falcon (Falco peregrinus) population in the Jura mountains 2000–2020. Ibis 164:217–239. https://doi.org/10.1111/ibi.12999

Kirkwood JK (1980) Management of a colony of common kestrels (Falco tinnunculus) in captivity. Lab Anim 14:313–316

Kjellén N, Roos G (2000) Population trends in Swedish raptors demonstrated by migration counts at Falsterbo, Sweden 1942–97. Bird Study 47:195–211. https://doi.org/10.1080/00063650009461175

Kleinstäuber G, Kirmse W, Langgemach T (2018) Nesting habitat selection of Peregrine Falcons (*Falco p. peregrinus*) in eastern Germany–the state of knowledge. Ornis Hung 26:259–273. https://doi.org/10.1515/orhu-2018-0034

Kokko H (1999) Competition for early arrival in migratory birds. J Anim Ecol 68:940–950. https://doi.org/10.1046/j.1365-2656.1999.00343.x

Konrad PM, Shaffer J, Igl LD (2020) The effects of management practices on grassland birds—Merlin (*Falco columbarius*). In: Johnson DH, Igl LD, Shaffer JA, DeLong JP (eds) The effects of management practices on grassland birds. U.S. Geological Survey Professional Paper 1842

Korpimäki E, Marti CD (1995) Geographical trends in trophic characteristics of mammal-eating and bird-eating raptors in Europe and North America. Auk 112:1004–1023. https://doi.org/10.2307/4089031

Korpimäki E, Norrdahl K (1991) Numerical and functional responses of Kestrels, Short-eared Owls and Long-eared Owls to vole densities. Ecology 72:814–826. https://doi.org/10.2307/1940584

Koskimies P (2005) Action plan for the Gyrfalcon (*Falco rusticolus*) in Europe. In Koskimies P, Lapshin NV (eds) Status of raptor populations in eastern Fennoscandia: proceedings of the workshop Nov. 8–10, 2005. Karelian Research Centre of the Russian Academy of Science Finnish-Russian Working Group on Nature Conservation, Kostomuksha, pp 70–79

Kovács A, Williams NP, Galbraith CA (2014) Saker Falcon *Falco cherrug* Global Action Plan (SakerGAP), including a management and monitoring system, to conserve the species. In: CMS Raptors MoU Coordinating Unit, Abu Dhabi. CMS Technical Series No. XX, Bonn. http://www.cms.int/en/document/2nd-draft-saker-falcon-global-action-plan-sakergap

Kuusela S, Solonen T (1984) The growth of kestrel nestlings in southern Finland. Ann Zool Fenn 21:309–312

Lack D (1946) Competition for food by birds of prey. J Anim Ecol 15:123–129. https://doi.org/10.2307/1552

Lack D (1950) The breeding seasons of European birds. Ibis 92:288–316. https://doi.org/10.1111/j.1474-919X.1950.tb01753.x

Landsborough Thomson A (1958) The migration of British Falcons (*Falconidae*) as shown by ringing results. Br Birds 51:179–188

Lazarova I, Petrov R, Andonova Y, Klisurov I, Dixon A (2021) Re-introduction of the Saker Falcon (*Falco cherrug*) in Bulgaria—preliminary results from the ongoing establishment phase by 2020. Biodivers Data J 9:e63729. https://doi.org/10.3897/BDJ.9.e63729

Leonardi G (2001) Lanner falcon. Birds Western Palearctic Update 3:161–178

Leonardi G (2015) The Lanner falcon. GLE, Catania

Leonardi G, Lopes L, Gallo Orsi U, Javed S, McGrady M (2024a) International Single Species Action Plan for the Sooty Falcon 2024–2036. CMS Raptors MOU, Abu Dhabi

Leonardi G, Sutton LJ, Alivizatos H, Andreotti A, Aktay-Sözüer L, Bacak E, Bilgin S, Binothman A et al (2024b) Reconstructing the paleobiogeographic history of the lanner falcon (*Falco biarmicus*) in the Mediterranean basin: towards a future perspective for its conservation. Ornithol Res 32:266–279. https://doi.org/10.1007/s43388-024-00196-y

Limiñana R, Romero M, Mellone U, Urios V (2012) Mapping the migratory routes and wintering areas of Lesser Kestrels Falco naumanni: new insights from satellite telemetry. Ibis 154:389–399. https://doi.org/10.1111/j.1474-919X.2011.01210.x

Lindberg P (1977) The Swedish Peregrine (*Falco peregrinus*) Project 1972–1976. In: Lindberg P (ed) Pilgrimsfalk. Report from a peregrine conference held at Grimsø wildlife research station, Sweden 1–2 April 1977. Swedish Society for the Conservation of Nature, Stockholm, pp 7–15

Lindberg P, Schei PJ, Wikman M (1988) The peregrine falcon in Fennoscandia. In: Cade TJ, Enderson JH, Thelander CG, White CM (eds) Peregrine falcon populations: their management and recovery. The Peregrine Fund, Boise, pp 159–172

Lovari S (1974) The feeding habits of four raptors in central Italy. Rap Res 8:45–57

Lundberg P (1988) The evolution of partial migration in birds. Trends Ecol Evol 3:172–175

Mak B (2024) On the tail of two species: the changing place of the peregrine falcon in society. People Nature 6:2157–2173. https://doi.org/10.1002/pan3.10710

Masman D, Daan S, Dijkstra C (1988) Time allocation in the kestrel (*Falco tinnunculus*), and the principle of energy minimization. J Anim Ecol 57:411–432. https://doi.org/10.2307/4914

Masman D, Dijkstra C, Daan S, Bult A (1989) Energetic limitation of avian parental effort: field experiments in the kestrel (*Falco tinnunculus*). J Evol Biol 2:435–455. https://doi.org/10.1046/j.1420-9101.1989.2060435.x

Matz A, Swem T, Johnson P et al (2011) Potential for climate change to increase contaminants exposure and effects in gyrfalcons. In: Watson RT, Cade TJ, Fuller M, Hunt G, Potapov E (eds) Gyrfalcons and Ptarmigan in a changing world. The Peregrine Fund, Boise, pp 161–175

McClure CJW, Westrip JRS, Johnson JA, Schulwitz SE, Virani MZ, Davies R et al (2018) State of the world's raptors: distributions, threats, and conservation recommendations. Biol Conserv 227:390–402. https://doi.org/10.1016/j.biocon.2018.08.012

McClure CJW, Schulwitz SE, Anderson DL, Robinson BW, Mojica EK, Therrien J-F, Oleyar MD, Johnson J (2019) Commentary: defining raptors and birds of prey. J Raptor Res 53:419–430. https://doi.org/10.3356/0892-1016-53.4.419

McCormack JE, Harvey MG, Faircloth BC, Crawford NG, Glenn TC, Brumfield RT (2013) A phylogeny of birds based on over 1500 loci collected by target enrichment and high-through put sequencing. PLoS One 8:e54848. https://doi.org/10.1371/journal.pone.0054848

McFadzen ME, Marzluff JM (1996) Mortality of prairie falcons during the fledging-dependence period. Condor 98:791–800. https://doi.org/10.2307/1369859

McGrady MJ, Al Fazari W, Al Jahdhami MH, Oli MK (2016) Survival of sooty falcons (*Falco concolor*) in Oman. J Ornithol 157:427–437. https://doi.org/10.1007/s10336-015-1302-6

McGrady MJ, Hines JE, Rollie CJ, Smith GD, Morton ER, Moore JF, Mearns RM, Newton I, Murillo-García OE, Oli MK (2017) Territory occupancy and breeding success of Peregrine Falcons *Falco peregrinus* at various stages of population recovery. Ibis 159:285–296. https://doi.org/10.1111/ibi.12443

Mearns R, Newton I (1984) Turnover and dispersal in a peregrine *Falco peregrinus* population. Ibis 126:347–355. https://doi.org/10.1111/j.1474-919X.1984.tb00255.x

Meijer THM (1989) Photoperiodic control of reproduction and molt in the kestrel, *Falco tinnunculus*. J Biol Rhythm 4:351–364. https://doi.org/10.1177/074873048900400304

Meijer T, Masman D, Daan S (1989) Energetics of reproduction in female kestrels. Auk 106:549–559. https://doi.org/10.1093/auk/106.4.549

Mellone U, López-López P, Limiñana R, Piasevoli G, Urios V (2013) The trans-equatorial loop migration system of Eleonora's falcon: differences in migration patterns between age classes, regions and seasons. J Avian Biol 44:417–426. https://doi.org/10.1111/j.1600-048X.2013.00139.x

Meyer SK, Spaar R, Bruderer B (2000) To cross the sea or to follow the coast? Flight directions and behaviour of migrating raptors approaching the Mediterranean Sea in autumn. Behaviour 137:379–399

Movalli P, Valvo ML, Pereira MG, Osborn D (2008) Organochlorine pesticides and polychlorinated biphenyl congeners in lanner Falco biarmicus feldeggii Schlegel chicks and lanner prey in Sicily, Italy. AMBIO Jour Hum Environ 37:445–452. https://doi.org/10.1579/0044-7447(2008)37[445:OPAPBC]2.0.CO;2

Negro JJ, Hiraldo F (1993) Nest-site selection and breeding success in the lesser kestrel *Falco naumanni*. Bird Study 40:115–119. https://doi.org/10.1080/00063659309477136

Negro JJ, De la Riva M, Bustamante J (1991) Patterns of winter distribution and abundance of lesser kestrels (*Falco naumanni*) in Spain. J Raptor Res 25:30–35

Negro JJ, Hiraldo F, Donázar JA (1997) Causes of natal dispersal in the lesser kestrel: inbreeding avoidance or resource competition? J Anim Ecol 66:640–648. https://doi.org/10.2307/5917

Newton I (1979) Population ecology of raptors. T & A D Poyser Ltd, Berkhamsted

Newton I, Wyllie I, Dale L (1999) Trends in the numbers and mortality patterns of Sparrowhawks (Accipiter nisus) and Kestrels (Falco tinnunculus) in Britain, as revealed by carcass analyses. J Zool Lond 247:139–147. https://doi.org/10.1111/j.1469-7998.1999.tb01190.x

Nielsen ÓK, Cade TJ (1990) Seasonal changes in food habits of Gyrfalcons in NE-Iceland. Ornis Scand 21:202–211. https://doi.org/10.2307/3676780

Nielsen BP, Christensen S (1970) Observations on the autumn migration of raptors in the Lebanon. Ornis Scand 1:65–73. https://doi.org/10.2307/3676025

Nielsen ÓK, Pétursson G (1995) Population fluctuations of gyrfalcon and rock ptarmigan: analysis of export figures from Iceland. Wildl Biol 1:65–71. https://doi.org/10.2981/wlb.1995.0011

Nilsson IN, Nilsson SG, Sylvén M (1982) Diet choice, resource depression and the regular nest spacing of birds of prey. Biol J Linn Soc 18:1–9. https://doi.org/10.1111/j.1095-8312.1982.tb02030.x

Nisbet ICT (1988) The relative importance of DDE and dieldrin in the decline of peregrine falcon populations. In: Cade TJ, Enderson JH, Thelander CG, White CM (eds) Peregrine falcon populations. Their management and recovery. The Peregrine Fund, Inc, Boise, pp 351–375

Norrdahl K, Korpimäki E (1993) Predation and interspecific competition in two Microtus voles. Oikos 66:149–158. https://doi.org/10.2307/3545105

Nygård T, Sandercock BK, Reinsborg T, Einvik K (2019) Population recovery of peregrine falcons in Central Norway in the 4 decades since the DDT-ban. Ecotoxicology 28:1160–1168. https://doi.org/10.1007/s10646-019-02111-4

Oli MK, Smith GD, McGrady MJ, Chaudhary V, Rollie CJ, Mearns R, Newton I, Lambin X (2023) Reproductive performance of Peregrine falcons relative to the use of organochlorine pesticides, 1946–2021. J Anim Ecol 92:2201–2213. https://doi.org/10.1111/1365-2656.14006

Orta J, Kirwan GM (2020) Lesser Kestrel (*Falco naumanni*), version 1.0. In: del Hoyo J, Elliott A, Sargatal J, Christie DA, de Juana E (eds) Birds of the world. Cornell Lab of Ornithology, Ithaca

Ortego J, Calabuig G, Aparicio JM, Cordero PJ (2008) Genetic consequences of natal dispersal in the colonial lesser kestrel. Mol Ecol 17:2051–2059. https://doi.org/10.1111/j.1365-294X.2008.03719.x

Palatitz P, Fehérvári P, Solt S, Horváth E (2015) Breeding population trends and pre-migration roost site survey of the Red-footed falcon in Hungary. Ornis Hung 23:77–93. https://doi.org/10.1515/orhu-2015-0007

Peakall DB (1975) PCB's and their environmental effects. CRC Crit Rev Environ Control 5:469–508

Peery ZM (2000) Factors affecting interspecies variation in home-range size of raptors. Auk 117:511–517. https://doi.org/10.1093/auk/117.2.511

Pescador M, Ramirez JIG, Peris SJ (2019) Effectiveness of a mitigation measure for the lesser kestrel (*Falco naumanni*) in wind farms in Spain. J Environ Manag 231:919–925. https://doi.org/10.1016/j.jenvman.2018.10.094

Pettifor RA (1983) The territorial behaviour of kestrels in arable fenland. Br Birds 76:206–214

Piross IS, Fehérvári P, Vas Z et al (2015) Louse (*Insecta*: *Phthiraptera*) infestations of the Amur falcon (*Falco amurensis*) and the red-footed falcon. Ornis Hung 23:58–65. https://doi.org/10.1515/orhu-2015-0005

Ponnikas S, Ollila T, Kvist L (2017) Turnover and post-bottleneck genetic structure in a recovering population of peregrine falcons Falco peregrinus. Ibis 159:311–323. https://doi.org/10.1111/ibi.12460

Porter R, Willis I (1968) The autumn migration of soaring birds at the Bosphorus. Ibis 110:520–536. https://doi.org/10.1111/j.1474-919X.1968.tb00061.x

Potters H (1998) Remarkably late onset of laying in Hobbies Falco subbuteo in 1997. De Takkeling 6:86–88

Prestt I (1965) An enquiry into the recent status of some of the smaller birds of prey and crows in Britain. Bird Study 12:196–221. https://doi.org/10.1080/00063656509476101

Price K (1947) The birds of Buckinghamshire. Records Buckinghamshire 15:20–31

Prommer M, Bagyura J, Chavko J, Uhrin M (2012) Migratory movements of Central and Eastern European Saker Falcons (*Falco cherrug*) from juvenile dispersal to adulthood. Aquila 119:111–134

Pyle P (2013) Evolutionary implications of synapomorphic wing-molt sequences among falcons (*Falconiformes*) and parrots (*Psittaciformes*). Condor 115:593–602. https://doi.org/10.1525/cond.2013.120173

Qninba A, Benhoussa A, El Agbani MA (2016) Oiseau de proie. A table avec le Faucon d'Eléonore. D'îles en îles 21:14–17

Quinn JL, Prop J, Kokorev Y, Black JM (2003) Predator protection or similar habitat selection in red-breasted goose nesting associations: extremes along a continuum. Anim Behav 65:297–307. https://doi.org/10.1006/anbe.2003.2063

Ratcliffe DA (1958) Broken eggs in peregrine eyries. Br Birds 51:23–26

Ratcliffe DA (1962) Breeding density in the peregrine *Falco peregrinus* and raven *Corvus corax*. Ibis 104:13–39. https://doi.org/10.1111/j.1474-919X.1962.tb08625.x

Ratcliffe DA (1963) The status of the Peregrine in Great Britain. Bird Study 10:56–90. https://doi.org/10.1080/00063656309476042

Ratcliffe DA (1967) Decrease in eggshell weight in certain birds of prey. Nature 215:208–210. https://doi.org/10.1038/215208a0

Ratcliffe DA (1970) Changes attributable to pesticides in egg breaking frequency and eggshell thickness in some British birds. J Appl Ecol 7:67–115. https://doi.org/10.2307/2401613

Ratcliffe DA (1972) The peregrine population of Great Britain in 1971. Bird Study 19:117–156. https://doi.org/10.1080/00063657209476336

Ratcliffe D (1993) The peregrine falcon, 2nd edn. T & AD Poyser, London

Rguibi H, Qninba A, Benhoussa A (2012) Eleonora's Falcon *Falco eleonorae*—updated state of knowledge and conservation of the nesting populations of the Mediterranean Small Islands. Albatross Project Monograph, Initiative PIM

Ristow D, Feldmann F, Scharlau W, Wink C, Wink M (1991) Population dynamics of Cory's shearwater (*Calonectris diomedea*) and Eleonora's falcon (*Falco eleonorae*) in eastern Mediterranean. In: Seitz A, Loeschcke V (eds) Species conservation: a population-biological approach. Birkhauser, Basel, pp 199–212

Rodríguez C, Bustamante J (2003) The effect of weather on lesser kestrel breeding success: can climate change explain historical population declines? J Anim Ecol 72:793–810. https://doi.org/10.1046/j.1365-2656.2003.00757.x

Rodríguez B, Siverio M, Rodríguez A, Siverio F (2007) Density, habitat selection and breeding success of an insular population of Barbary Falcon *Falco peregrinus pelegrinoides*. Ardea 95:213–223. https://doi.org/10.5253/078.095.0205

Rudebeck G (1950) Studies on Bird Migration. Vår Fågelvärld. Suppl. 1. Lund
Rudebeck G (1951) The choice of prey and modes of hunting of predatory birds with special reference to their selective effect. Oikos 3:200–231. https://doi.org/10.2307/3564663
Safriel U (1968) Bird migration at Elat, Israel. Ibis 110:283–320. https://doi.org/10.1111/j.1474-919X.1968.tb00039.x
Serrano D, Tella JL (2003) Dispersal within a spatially structured populations of lesser kestrels: the role of spatial isolation and conspecific attraction. J Anim Ecol 72:400–410. https://doi.org/10.1046/j.1365-2656.2003.00707.x
Serrano D, Tella JL (2012) Lifetime fitness correlates of natal dispersal in a colonial bird. J Anim Ecol 81:97–107. https://doi.org/10.1111/j.1365-2656.2011.01878.x
Serrano D, Tella JL, Forero MG, Donazar JA (2001) Factors affecting breeding dispersal in the facultatively colonial lesser kestrel: individual experience vs. conspecific cues. J Anim Ecol 70:568–578. https://doi.org/10.1046/j.1365-2656.2001.00512.x
Serrano D, Forero MG, Donázar JA, Tella JL (2004) Dispersal and social attraction effect colony selection and dynamics of lesser kestrels. Ecology 85:3438–3447. https://doi.org/10.1890/04-0463
Serrano D, Oro D, Ursúa E, Tella JL (2005) Colony size selection determines adult survival and dispersal preferences: Allee effects in a colonial bird. Am Nat 166:E22–E31. https://doi.org/10.1086/431255
Serrano D, Carrete M, Tella JL (2008) Describing dispersal under habitat constraints: a randomization approach in lesser kestrels. Basic Appl Ecol 9:771–778. https://doi.org/10.1016/j.baae.2007.08.013
Shrubb M (1982) The hunting behaviour of some farmland kestrels. Bird Study 29:121–128. https://doi.org/10.1080/00063658209476746
Sielicki S, Sielicki J (2009) Restoration of Peregrine falcon in Poland 1989-2007. In: Sielicki J, Mizera T (eds) Peregrine falcon populations - status and perspectives in the 21st century. Turul-Poznan University of Life Science Press, Poznan, pp 699–722
Simmons KEL (1951) Raptor migration in the Suez area. Ibis 93:402–406. https://doi.org/10.1111/j.1474-919X.1951.tb05442.x
Slagsvold T, Sonerud GA (2007) Prey size and ingestion rate in raptors: importance for sex roles and reversed sexual size dimorphism. J Avian Biol 38:650–666. https://doi.org/10.1111/j.2007.0908-8857.04022.x
Slobodník R, Chavko J, Lengyel J, Noga M, Maderič B, Baláž M (2017) Trend in an isolated population of the red-footed falcon (*Falco vespertinus*) at the edge of its breeding range (southwestern Slovakia). Slovak Raptor J 11:83–89. https://doi.org/10.1515/srj-2017-0007
Smith GD, Murillo-Garcia OE, Hostetler JA, Mearns R, Newton I, McGrady MJ, Oli MK (2015) Demography of population recovery: survival and fidelity of Peregrine Falcons at various stages of population recovery. Oecologia 178:391–401. https://doi.org/10.1007/s00442-014-3168-3
Smits J, Naidoo V (2018) Toxicology of birds of prey. In: Sarasola JH, Grande JM, Negro JJ (eds) Birds of prey. Biology and conservation in the XXI century. Springer, Cham, pp 229–250. https://doi.org/10.1007/978-3-319-73745-4_10
Snow DW (1968) Movement and mortality of British kestrels *Falco tinnunculus*. Bird Study 15:65–83. https://doi.org/10.1080/00063656809476184
Solonen T (1993) Spacing of birds of prey in southern Finland. Ornis Fennica 70:129–143
Spina F, Leonardi G (2007) Piano d'azione nazionale per il Falco della regina (*Falco eleonorae*). Quad Cons Natura. No. 26. Ministero dell'Ambiente e della Tutela del Territorio e del Mare, INFS, Rome
Strandberg R, Klaasen RHG, Hake M, Olofsson P, Alerstam T (2009a) Converging migration routes of Eurasian hobbies *Falco subbuteo* crossing the African equatorial rain forest. Proc R Soc B 276:727–733. https://doi.org/10.1098/rspb.2008.1202
Strandberg R, Klaasen RHG, Hake M, Olofsson P, Alerstam T (2009b) Daily travel schedules of adult Eurasian hobbies *Falco subbuteo*—variability in flight hours and migration speed along the route. Ardea 97(3):287–295. https://doi.org/10.5253/078.097.0304

Stresemann E (1927–1934) Aves. In: Kükenthal W, Krumbach T (eds) Handbuch der Zoologie. de Gruyter, Berlin/Leipzig

Stresemann E (1968) Der Eingrift der Eleonorenfalken in den herbstlichen Vogelzug. J Ornithol 109:472–474. https://doi.org/10.1007/BF01671583

Strucker R, Luitwieler M (2015) Late summer concentration of Hobbies *Falco subbuteo* on Goeree. De Takkeling 23:154–156

Suhonen J, Norrdahl K, Korpimäki E (1994) Avian predation risk modifies breeding bird community on a farmland area. Ecology 75:1626–1634. https://doi.org/10.2307/1939623

Svärdson G (1949) Competition and habitat selection in birds. Oikos 1:157–174. https://doi.org/10.2307/3564710

Terraube J, Vasko V, Korpimäki E (2015) Mechanisms and reproductive consequences of breeding dispersal in a specialist predator under temporally varying food conditions. Oikos 124:762–771. https://doi.org/10.1111/oik.01974

Thiollay JM (1968) Le régime alimentaire de nos rapaces: quelques analyses françaises. Nos Oiseaux 29:249–269

Thomasson K (1947) Något om pilgrimsfalkens boplatsval (On the nesting-sites of the peregrine falcon in the countries around the Baltic). Vår Fågelvärld 6:72–81

Thomson AL (1958) The migrations of British falcons (*Falconidae*) as shown by ringing results. Br Birds 51:179–188

Tolonen P, Korpimäki E (1994) Determinants of parental effort: a behavioural study in the Eurasian kestrel, *Falco tinnunculus*. Behav Ecol Sociobiol 35:355–362. https://doi.org/10.1007/BF00184424

Uttendörfer O (1939) Die Ernährung der deutschen Raubvögel und Eulen und ihre Bedeutung in der heimischen Natur. Neumann-Neudamm, Berlin

Valkama J, Korpimäki E (1999) Nest-box characteristics, habitat quality and reproductive success of Eurasian Kestrels. Bird Study 46:81–88. https://doi.org/10.1080/00063659909461117

van den Hout PJ, Spaans B, Piersma T (2008) Differential mortality of wintering shorebirds on the Banc d'Arguin, Mauritania, due to predation by large falcons. Ibis 150:219–230. https://doi.org/10.1111/j.1474-919X.2008.00785.x

Vasko V, Laaksonen T, Valkama J, Korpimäki E (2011) Breeding dispersal of Eurasian kestrels *Falco tinnunculus* under temporally fluctuating food abundance. J Avian Biol 42:552–563. https://doi.org/10.1111/j.1600-048X.2011.05351.x

Vaughan R (1961) *Falco eleonorae*. Ibis 103:114–130. https://doi.org/10.1111/j.1474-919X.1961.tb02424.x

Vaurie C (1961a) Systematic Notes on Palearctic Birds. No. 44 *Falconidae*: the genus Falco (Part 1, *Falco peregrinus* and *Falco pelegrinoides*). Am Mus Nov No. 2035, New York

Vaurie C (1961b) Systematic Notes on Palearctic Birds. No. 45 *Falconidae*: the genus *Falco* (Part 2). Am Mus Nov No. 2038, New York

Walker CH, Newton I (1999) Effects of cyclodiene insecticides on raptors in Britain- correcting and updating of an earlier paper by Walker and Newton, Ecotoxicology 7:185–189 (1998). Ecotoxicology 8:425–429. https://doi.org/10.1023/A:1008995400871

Walker LA, Beith SJ, Lawlor AJ, Moeckel C, Peréira MG, Potter ED, Shore RF (2011) Persistent Organic Pollutants (POPs) and inorganic elements in predatory bird livers and eggs 2007 to 2009: a Predatory Bird Monitoring Scheme (PBMS) Report. Centre for Ecology & Hydrology, Lancaster

Waller R (1962) Der Wilde Falk ist mein Gesell. Verl. Nermann-Neudamm, Melsungen

Walter H (1968) Zur Abhängigkeit des Eleonorenfalken (*Falco eleonorae*) vom mediterranen Vogelzug. J Ornithol 109:323–365. https://doi.org/10.1007/BF01678378

Wayre P, Joyy GF (1958) Notes on the breeding of the Iceland gyr falcon. Br Birds 51:285–290

Weaving A, Jackson HA, Nicholls MK, Franklin J, Vega R (2021) Conservation genetics of regionally extinct peregrine falcons (*Falco peregrinus*) and unassisted recovery without genetic bottleneck in southern England. Conserv Genet 22:133–150. https://doi.org/10.1007/s10592-020-01324-z

Wegner P, Kleinstäuber G, Baum F, Schilling F (2005) Long-term investigation of the degree of exposure of German peregrine falcons (*Falco peregrinus*) to damaging chemicals from the environment. J Ornithol 146:34–54. https://doi.org/10.1007/s10336-004-0053-6

White CM, Sonsthagen SA, Sage GK, Anderson C, Talbot SL (2013) Genetic relationships among some subspecies of the Peregrine Falcon (*Falco peregrinus* L.), inferred from mitochondrial DNA control-region sequences. Auk 130:78–87. https://doi.org/10.1525/auk.2012.11173

Wiebe LK, Jönson KI, Wiehn J, Hakkarainen H, Korpimäki E (2000) Behaviour of female Eurasian kestrels during laying: are there time constraints on incubation? Ornis Fenn 77:1–9

Wiehn J, Ilmonen P, Korpimäki E, Pahkala M, Wiebe KL (2000) Hatching asynchrony in the Eurasian kestrel *Falco tinnunculus*: an experimental test of the brood reduction hypothesis. J Anim Ecol 69:85–95. https://doi.org/10.1046/j.1365-2656.2000.00374.x

Wightman CS, Fuller MR (2006) Influence of habitat heterogeneity on distribution, occupancy patterns, and productivity of breeding Peregrine Falcons in central West Greenland. Condor 108:270–281. https://doi.org/10.1093/condor/108.2.270

Wiklund CG (1979) Increased breeding success for Merlins *Falco columbarius*, nesting among colonies of Fieldfares *Turdus pilaris*. Ibis 121:109–111. https://doi.org/10.1111/j.1474-919X.1979.tb05026.x

Wiklund CG (1990) The adaptive significance of nest defence by merlin, *Falco columbarius*, males. Anim Behav 40:244–253. https://doi.org/10.1016/S0003-3472(05)80919-4

Wiklund CG (1996) Determinants of dispersal in breeding Merlins *Falco columbarius*. Ecology 77:1920–1927. https://doi.org/10.2307/2265795

Wiklund C (2001) Food as a mechanism of density-dependent regulation of breeding numbers in the merlin, *Falco columbarius*. Ecology 82:860–867. https://doi.org/10.1890/0012-9658(2001)082[0860:FAAMOD]2.0.CO;2

Wiklund CG, Larsson BL (1994) The distribution of breeding Merlins *Falco columbarius* in relation to food and nest sites. Ornis Svecica 4:113–122

Wiklund CG, Village A (1992) Sexual and seasonal variation in territorial behaviour of kestrels, *Falco tinnunculus*. Anim Behav 43:823–830. https://doi.org/10.1016/S0003-3472(05)80205-2

Wille F (1977) The Peregrine (*Falco peregrinus*) in Denmark. In: Lindberg P (ed) Pilgrimsfalk. Report from a peregrine conference held at Grimsø wildlife research station, Sweden 1–2. April 1977. Swedish Society for the Conservation of Nature, Stockholm, p 31

Wilson MW, Balmer DE, Jones K, King V.A, Raw D, Rollie CJ, Rooney E, Ruddock M, Smith GD, Stevenson A, Stirling- Aird PK, Wernham CV, Weston JM, Noble DG (2018). The breeding population of Peregrine Falcon *Falco peregrinus* in the United Kingdom, Isle of Man and Channel Islands in 2014. Bird Study 65:1–19. https://doi.org/10.1080/00063657.2017.1421610

Wink M, Seibold I (1996) Molecular phylogeny of Mediterranean raptors (Families Accipitridae and Falconidae). Biol Conserv Mediter Raptors Monogr 4:335–44

Wrege PH, Cade TJ (1977) Courtship behaviour of large falcons in captivity. Rap Res 11:1–46

Xirouchakis S, Fric J, Kassara C, Portolou D, Dimalexis A, Karris G et al (2012) Variation in breeding parameters of Eleonora's falcon (*Falco eleonorae*) and factors affecting its reproductive performance. Ecol Res 27:407–416. https://doi.org/10.1007/s11284-011-0912-6

Yaneva S, Gradev G, Bileva T (2022) Different types of nest boxes used by Lesser Kestrel (*Falco naumanni*) after being recovery as a breeder in Bulgaria. Sci Papers Ser D Anim Sci 45:664–669

Yuri T, Kimball RT, Harshman J et al (2013) Parsimony and model-based analyses of indels in avian nuclear genes reveal congruent and incongruent phylogenetic signals. Biology 2(1):419–444. https://doi.org/10.3390/biology2010419

Zabala J, Zuberogoitia I (2015) Breeding performance and survival in the peregrine falcon *Falco peregrinus* support an age-related competence improvement hypothesis mediated via an age threshold. J Avian Biol 45:141–150. https://doi.org/10.1111/jav.00505

Zellweger-Fischer J, Schaub M, Müller C, Rudin M, Spiess M, Jenni L (2011) Der Bruterfolg des Turmfalken *Falco tinnunculus*: resultate und Erkenntnisse aus fünf Jahren integriertem Populationsmonitoring. Ornithol Beob 108:37–54

Zinevich L, Prommer M, Laczkó L, Rozhkova D, Sorokin A, Karyakin I, Bagyura J, Cserkész T, Sramkó G (2023) Phylogenomic insights into the polyphyletic nature of Altai falcons within eastern sakers (*Falco cherrug*) and the origins of gyrfalcons (*Falco rusticolus*). Sci Rep 13(17800). https://doi.org/10.1038/s41598-023-44534-4

Żmihorski M, Rejt Ł (2007) Weather-dependent variation in the cold-season diet of urban kestrels *Falco tinnunculus*. Acta Ornithol 42:107–113. https://doi.org/10.3161/000164507781646861

Zuberogoitia I, Zabala J, Martínez JE, Olsen J (2015) Alternative eyrie use in peregrine falcons: is it a female choice? J Zool 296:6–14. https://doi.org/10.1111/jzo.12221

Index

G. Leonardi, *The History of Western Palearctic Falcons*,
https://doi.org/10.1007/978-3-031-86528-2

The manufacturer's authorised representative in the EU is Springer Nature Customer Service Centre GmbH, Europaplatz 3, 69115 Heidelberg, Germany. If you have any concerns regarding our products, please contact ProductSafety@springernature.com

Printed and bound by CPI Group (UK) Ltd, Croydon, CR0 4YY

30/09/2026

02225584-0003